D1534819

This item no longer
belongs to Davenport
Public Library

PRAISE FOR

Carving Out a Living on the Land

"Resourcefulness, inventiveness, a sense of tradition, a relentless capacity for hard physical labor—these are the qualities that surface again and again in *Carving Out A Living On the Land*."

— From the foreword by **Verlyn Klinkenborg**

"In this no-nonsense resource for novice Christmas Tree farmers (and all farmers), Van Driesche shares his sustainable success story in pragmatic and engaging detail."

— **Forrest Pritchard**, farmer; *New York Times* bestselling author of *Gaining Ground*

"A couple of years ago I bought a Christmas tree from Emmet Van Driesche's little family in the woods. I immediately wanted to move in with them. They radiated the kind of peaceful contentment that comes with doing work that makes sense to you and that you are good at. But since moving in with them would be creepy, I am glad to have this book, a thorough and generous guide to shaping your world not only to your own happiness, but the world's as well."

— **John Hodgman**, host of *Judge John Hodgman*; author of *Vacationland*

"*Carving Out a Living on the Land* describes the joy of living a simpler life in tune with nature, without skipping on the nitty gritty of hard work and paying the bills. Along the way, Emmet shows that spoon carving is not just a wonderful pastime but can be part of a thriving modern business, filling one's time between the hustle and bustle of childcare and tree cutting with calm, spoon-y industriousness."

— **Barn the Spoon**, founder, Spoon Club and The Green Wood Guild; author of *Spoon*

Carving Out
a Living
on the Land

DAVENPORT PUBLIC LIBRARY
321 MAIN STREET
DAVENPORT, IOWA 52801

Carving Out a Living on the Land

Lessons in Resourcefulness and Craft from an Unusual Christmas Tree Farm

Emmet Van Driesche

Foreword by VERLYN KLINKENBORG

Chelsea Green Publishing
White River Junction, Vermont
London, UK

Copyright © 2019 by Emmet Van Driesche.
All rights reserved.

Unless otherwise noted, all photographs copyright © 2019 by Emmet Van Driesche.

No part of this book may be transmitted or reproduced in any form by any means without permission in writing from the publisher.

Project Manager: Alexander Bullett
Editor: Michael Metivier
Copy Editor: Deborah Heimann
Proofreader: Laura Jorstad
Indexer: Margaret Holloway
Designer: Melissa Jacobson

Printed in Canada.
First printing May 2019.
10 9 8 7 6 5 4 3 2 1   19 20 21 22 23

Our Commitment to Green Publishing

Chelsea Green sees publishing as a tool for cultural change and ecological stewardship. We strive to align our book manufacturing practices with our editorial mission and to reduce the impact of our business enterprise in the environment. We print our books and catalogs on chlorine-free recycled paper, using vegetable-based inks whenever possible. This book may cost slightly more because it was printed on paper that contains recycled fiber, and we hope you'll agree that it's worth it. Chelsea Green is a member of the Green Press Initiative (www.greenpressinitiative.org), a nonprofit coalition of publishers, manufacturers, and authors working to protect the world's endangered forests and conserve natural resources. *Carving Out a Living on the Land* was printed on paper supplied by Marquis that is made of recycled materials and other controlled sources.

Library of Congress Cataloging-in-Publication Data
Names: Van Driesche, Emmet, author.
Title: Carving out a living on the land : lessons in resourcefulness and craft from an unusual
    Christmas tree farm / Emmet Van Driesche.
Description: White River Junction, Vermont : Chelsea Green Publishing, [2019]
    | Includes bibliographical references and index.
Identifiers: LCCN 2019000401| ISBN 9781603588263 (hardcover : alk. paper)
    | ISBN 9781603588270 (ebook)
Subjects: LCSH: Christmas tree growing—Massachusetts. | Farmers—Massachusetts—Biography.
Classification: LCC SB428.3 .V36 2019 | DDC 635.9/775—dc23
LC record available at https://lccn.loc.gov/2019000401

Chelsea Green Publishing
85 North Main Street, Suite 120
White River Junction, VT 05001
(802) 295-6300
www.chelseagreen.com

*For Cecilia, with me shoulder-to-shoulder;
and for Al Pieropan, who started it all.*

CONTENTS

FOREWORD

I don't know where Emmet Van Driesche learned to write, but I know that I had nothing to do with it. Fifteen years ago, in the fall of 2004, he took a nonfiction writing seminar I was teaching at Bard College. I recently re-read all the work Emmet submitted that semester, and I can't imagine what I found to say about it, except, perhaps, "More, please." When a new semester begins, I always have a recurring fear: What if the students write so well that I have nothing to teach them? My fear is a class full of Emmets.

Too many Emmets isn't a problem most people have to worry about. In fact judging by the book you hold in your hands, most of us would probably welcome many Emmets—on our farms, in our gardens, in our communities, in our conversations about the future. Resourcefulness, inventiveness, a sense of tradition, a relentless capacity for hard physical labor—these are the qualities that surface again and again in *Carving Out A Living On the Land*, along with a profound awareness of a shared ecological imperative.

That Emmet happens also to write well might seem like nothing more than a pleasant surprise. After all, you don't have to be a good writer to be a good farmer or a far-thinking environmentalist. It wouldn't surprise me to learn that in a list of desirable agricultural attributes most farmers would put "good writer" way below "good pinochle player"—and miles and miles behind, say, "good hydraulic mechanic."

But if you're going to write about your farm, it helps to be a good writer. Every good book about farming inevitably proposes an argument about the nature of good farming, and the clearer that argument is, the better. And at its root, every good farm is also a work of imagination, a sustained exercise in problem-solving. Being a good farmer is more than an occupation: It's a preoccupation. In these and other ways, the good farmer resembles the good writer.

And there's a deeper link between the farmer and the writer, a kinship you can find in two very different essays: Aldo Leopold's "The Farmer as Conservationist" from 1939 and George Orwell's "Politics and the English Language" from 1946. At first these essays, like their authors, seem to have no connection. Orwell is writing about the corruption of language by politics, and Leopold is writing about loss of diversity and complexity in the agricultural landscape. But both make the underlying argument that only individuals—not institutions—can make a difference, and, to shift the emphasis slightly, that individuals *can* make a difference. The vitality of conservation—in language and landscape—relies on what Leopold calls "a positive exercise of skill and insight, not merely a negative exercise of abstinence and caution." As Orwell says, looking over the ravages of the early twentieth century, "one cannot change this all in a moment, but one can at least change one's own habits."

I don't mean to load the arguments of these remarkable essays onto Emmet's shoulders. He has enough to carry as it is, great bundles of balsam trimmings not least among them. Nor is *Carving Out a Living on the Land* polemical: It isn't making a grand proposition or telling you how to live. It simply shows us the choices that Emmet and his wife, Cecilia, have made and how they have made them. In a sense they are the new settlers—not lighting out for the territory or hoping to find virgin ground but taking over the land and, to a large extent, the practices of a local Christmas tree farmer in rural Massachusetts. If there's a treatise hidden in this book, it concerns the way the economies of the land and money itself are overlaid on the economy of the farmer's body.

At the heart of Emmet's farming practice is the venerable method of coppicing—the regrowth of tree shoots from a stump (or stool, to use the proper word). Picture a Christmas tree farm in your head, and what you're likely to be picturing is a short, regimented forest of young trees—a plantation—waiting to be cut down one by one as they come to maturity, with no further use for what's left behind. What Emmet relies upon instead is the exuberance of the tree itself, its willingness to regrow and to be reharvested season after season.

As I read *Carving Out a Living in the Land*, I found myself looking forward to the sentences that begin, "I'm not an expert . . ." Those sentences are nearly always followed by a fund of common sense, the kind of things you learn only by trial and error. You begin to realize that throughout this book Emmet is coppicing his own mind: Where one idea gets cut down, two more spring into life. It's impossible to watch this happening without feeling a similar regeneration in the plantations of one's own head.

—Verlyn Klinkenborg
East Chatham, New York

Introduction

Thehe air is cold enough for my breath to show, but I'm about to break a sweat. I'm harvesting balsam branches, grabbing each with one hand and cutting them with the red clippers in the other. I grasp each new branch with the same hand until my fingers can't stretch around any more, then shove my arm through the middle of the entire handful and keep going. I work fast and don't stop until my arm is completely stacked with branches and sticking straight out, and I look like a kid with too many sweaters on under his jacket. Pivoting on my heel, I stride back to my central pile of balsam boughs and dump the armload on top, eyeballing it to gauge how much the pile weighs. I decide I need more and head off in another direction into the grove.

The balsam fir grows from big, wild stumps, in thickets that can stretch 20 feet around, the trees crowded so closely together, in no apparent order or pattern, that their branches interlock. Instead of single trees, each stump has up to three small trees of different ages growing off it. They are pruned as Christmas trees, and I am a Christmas tree farmer.

My Christmas tree farm is unlike any you have ever seen or imagined, however. For one thing, my trees do not march in rows across a field, but rather spring out of the ground in loose affiliations knit together by a lacework of paths. But the biggest difference is that my trees are coppiced, growing out of the top or sides of a stump that has a thick skirt of branches. These branches keep the stump alive; each year the stump sends out dozens of new shoots, each one vying to be the new tree. Unlike conventional Christmas tree farms, my problem isn't buying enough seedlings to meet demand ten years down the road, or keeping newly planted stock alive through a summer drought. My problem is abundance, the sheer exuberance of nature that tries to produce a dozen trees where I want just one. Shoots erupt from the rim of the stump, from the bark on the sides, and from the tops of branches. Sometimes whole branches start to curve up and make a bid for the sky.

Carving Out a Living on the Land is a practical book about a specific farm, but it's not a blueprint for other farmers, Christmas tree farmers or otherwise. Our farm is the result of the combination of the circumstances my family found ourselves in, and what we've made of those opportunities. Under different circumstances we would have made different choices. Maybe those would have still led us to this place, but maybe not. What *is* certain is that we each inhabit our own reality and bring our own strengths to the table. Your choices will be yours. Your outcome will look like you.

This book examines what it means to make land productive, and how you can combine your interests with what your land can produce to piece together a living. In one sense, this is nothing new; it is the story of every farm that is, was, and ever will be. But in the broadest definition, a farm is the intersection of the land's potential to produce and a market's potential for demand. All farmers are constantly figuring out the angles that best combine what they know how to do, what they have the land or resources to do, what will make them the most money, and what their market wants. To some extent, a farm's prosperity is a good indication of how well these four things mesh in that particular combination, and the combinations are endless. Most businesses are short lived, and farms are no exception. They fail either quickly because they don't find the right combination of these factors, or after existing for a long time because they don't adjust when some of these factors change. Even if you seek out a particular piece of land or business opportunity with a given outcome in mind, once you make that initial plunge, farming becomes reactive. The process of observing and responding to reality continues for the rest of your life.

Introduction

I do unapologetically advocate for a certain approach to farming in this book, one that values flexibility and diversity over scale. I will admit that this has been our path because of the constraints of our situation, but all farms exist on a spectrum of how well their current form supports the financial, emotional, and creative needs of the farmers. A typical New England dairy farm, like so many holding on in my area, is at one end of the spectrum, pressed thin under the weight of a century of barns and outbuildings in need of repair, producing commodity milk that limits the farmer's control of their own income, and bearing both high overhead costs and high risk. At the other end, I would argue, is our farm: 10 acres of coppiced Christmas trees that for sixty years have quietly been producing income with very little in the way of costs and with a great degree of flexibility in pricing and supplemental streams of income.

I began writing this book at first simply to share what an amazing place this farm is, and to make the techniques it employs more broadly known to the farming community. But as it progressed, it became clear that what our story has to offer is much broader and more important than how to grow Christmas trees this way or how to tie a balsam wreath. Our farm is a living example of how to make a landscape economically productive while at the same time respecting the function of a complex ecology. It is an old way of doing things, persisting for so long that it has come back around to feeling new and experimental. My hope is that this book will inspire you to look at your land (or land you hope to purchase) in a different light. Instead of dividing land into "good" field or pasture and all the rest, I want you to think about what might be done with that rocky section, or ditch, or patch of trees too small to log. What could you grow there? How would the business side of that work? Equally important, how can you leverage your skills and knowledge and opportunities from the farm to create income that is not tied so directly to the land? Maybe that's teaching. Maybe consulting. Maybe mastering a craft or starting a supply business or even, if you are really crazy, writing a book. Farming used to be something that almost everyone did, and almost all farmers also had other work to make ends meet. Only in the past hundred years have we developed the rigid notion that if you are a farmer you should only farm, or that if you have a source of income other than farming you are somehow not really farming. Leveraging your skills to provide income separate from the farm is a key component to keeping the farm (and your own career and mental state) healthy.

If you live in the city and just enjoy thinking about these things, that's great. If you dream someday of buying or leasing a bit of land and doing your own thing, there is lots in here for you. If you already have land of any kind, this book will provide guideposts (not a blueprint, not a map, more like the rock cairns that guide you to mountain summits above the tree line) touching on the specifics of how to take a dream and make it into a real business. One customer at a time. One sale at a time. One day at a time.

It is the nature of a managed forest to overproduce, and my job is to cull and thin and push it back until it produces just the right amount for my goals. With careful tending, a single stump can have multiple trees growing out of it at any given time: one fully grown, one halfway, and a couple of sprouts competing to see which will prove the best. Each stump produces a harvestable tree about every four years, and about every four years its skirt of branches can be harvested and regrown.

I keep adding armloads of branches to the pile until I judge it to be about 50 pounds (23 kg), then bundle it up with a length of baling twine. There are tricks to tying a bale of balsam properly. If you just pull on the twine as it comes through the loop at one end to cinch it tight, the jute fibers cut right through themselves. Instead, you need to kneel on the boughs, pressing them tight with your body, and then flip them on their side and press down further with your knee as you gently tighten the twine and knot it off. It is not unlike how a shearer manipulates a sheep, with pressure from the knees and legs and an implacable suppleness. It is also a decent metaphor for how to make a successful business: knowing where to squeeze and how to gather up the slack.

Once the bale is tied up I flip it onto my back, where it rides at the base of my neck. I stoop forward under the weight and have to tilt my head to the side to see. Thankfully, I know where I'm going, and I navigate my way down the slope by a hundred tiny landmarks that I know intimately after eight years of doing this. When I first started working this grove I got lost all the time. Even though I knew that if I

The view with a bale of white pine boughs on my back.

headed downhill I would eventually come out on the dirt road, I was continually surprised at where a given path led me. Now I know these paths better than the lines on my palm. I know them even as they change from year to year as big trees get cut and little trees grow big. They are a part of me. I have built and maintained them and walked them in every mood and state of exhaustion.

The trail I am on flows into a larger one, a main artery coursing downslope to the road. Although it's cold, there is no snow on the ground, and the patchwork quilt of mosses, grasses, and understory plants muffles my footsteps. Just as the trees are not in rows, they are also not growing in a grassy meadow. This is truly a managed forest, and the thousands of deciduous saplings springing up everywhere remind me that the forest is always just a couple years of idleness away from reclaiming its sovereignty. The ground is rocky, with glacial erratics hiding in the undergrowth and jutting out of the trails. Rhododendrons, azaleas, huckleberries, and willow thickets shoulder their way in for some sunlight. Some of these species have economic value—the winterberry I use to decorate wreaths, the willow I plan to wholesale to basket makers, the trees I harvest to carve into spoons, and the saplings I cut to make scythe handles—but many species, while they have no direct value, still make up an ecosystem that is resilient because of this very complexity.

Undecorated wreaths are stored on poles to keep them from being squashed flat.

I have been harvesting greens about as far away from the road as possible, so it takes five minutes to walk down, and by the time I hit the gravel my back is aching and my arms, reaching over my head to stabilize the load, are numb. This particular bale definitely weighs more than 50 pounds. I don't always judge bale weight accurately, and when I have such a long walk to haul each one down, I am tempted to keep throwing boughs on the pile in a foolish bid to make each trip more efficient. I come down to the road in a flurry of fast steps, turn right, and trudge up to the hut and to the tarp structure where I keep my stash of bales cool and out of the sun.

The hut is tiny and shingled, built on a trailer frame and parked right next to the road. Behind it and to one side is a framework of logs supporting a large gray canvas tarp. Its shape is irregular because the posts are living trees, and I had to work with what was available. The ridgeline of the tarp barn is a good 12 feet (3.5 m) off the ground, a 20-foot-long (6 m) ash log bolted between two birch trees. Ducking under the low edge of the tarp, I tip the bale of greens over my head and onto the top of a very large pile of similar bundles. I am running out of room under the tarp, with bales of greens nearly touching the fabric in a windrow extending 20 feet long, 6 feet (1.8 m) deep, and 6 high. The other half of the structure is occupied by a series of 6-foot-long poles, one end supported on a log beam and the other suspended from the ridgepole on loops of baling twine. Each of these poles supports ten to fifteen wreaths, depending on the wreath size. I have fifteen poles full so far and only a few more spots left.

I've prepared this stockpile for the rush of wholesale orders that comes right before Thanksgiving each year, as businesses prepare to switch to Christmas mode on Black Friday. Very soon I will start delivering load after load of greens and wreaths and trees to garden centers and farm stores and co-ops within an hour-and-a-half radius, and at that point the limiting factor will no longer be storage, but rather my time. By Thanksgiving, these piles will be gone and I will be coming to the farm extra early to cut even more bales in case a last-minute order comes in. It always does.

It is the nature of a seasonal business to have external limits. Christmas comes when it comes, and you either take advantage of the demand or you don't. Less obvious are the internal limits, those of

Just a portion of the amphitheater of trees that makes up the 5-acre You-Cut grove.

yourself and your life, or those of the land. I only have so many hours of daylight. My crew likewise only wants to work a certain amount, and I respect that. Even less elastic is the limit of what the land will sustainably produce year after year. Trees are in the most limited supply, followed by greens and then wreaths. Each year I make a guess as to when I should say no to a big order, sometimes because I don't have the time but more often because I have a hunch that we have reached the limit of what the land will bear for now.

Ducking back out from the tarp barn, I walk over to the hut and enter a tiny workspace, 7 × 12 feet (2 × 3.5 m), with a small woodstove in the corner opposite the door and a workbench spanning one entire end. This is the hub of operations, with pruning saws in leather scabbards hanging on the wall and clippers piled on the bench. Bundles of wire hoops and bushel baskets of pinecones occupy the corners while rolls of red ribbons are stacked on a shelf up high. I shed my coat and hang it on a hook, then reach down and start pulling baling twine from a spool at my feet. I pull out three arm spans' worth, cut it with clippers, and pivot toward the door, coiling the twine on my fingers as I walk. I am moving fast enough that the tail end of the twine slips out the door just before it swings shut. I've got work to do.

CHAPTER 1

Your Starting Place

In the fall of 2008, my wife, Cecilia, and I moved into the back half of a ramshackle farmhouse perched right where the road switched from paved to dirt after curving around two large barns in varying states of repair and a number of smaller outbuildings. The house was blue-gray, with white trim and stone porches. The views from our apartment were sweeping, looking up at a meadow studded by a pair of dying apple trees.

We had been living and working on a small vegetable farm five minutes down the road in the same rural Massachusetts town, but had made a decision to have kids hand in hand with a decision to stop farming—and to move out of the 8 × 16 foot (2.5 × 5 m) tiny house we'd built and had been living in for eight months. We wanted, to take the advice of one farmer my wife knew up in Maine, where she had farmed for years, to try as hard as we could *not* to farm. If we couldn't not farm, then we would farm.

Our new home was scrappy and worn at the heels, but we loved it. Our landlord was happy to let us paint the rooms any colors we wanted, and over the next seven years we took over much of the maintenance and repair of the building, cleaning the chimneys, patching the slate roof, fixing broken windows, mucking out and insulating the basement, dealing with plumbing leaks, and all the usual tasks associated with homeownership that renters normally miss out on. It would prove to be an invaluable learning experience.

Our landlord was Al Pieropan, who with his wife, Mimi, bought the farmhouse and 25 acres in 1953, and in the ensuing fifty-five years built two more houses on the property: another rental and the house they lived in. Al also planted 10 of the acres to Christmas trees over the course of twenty years, using seedlings yanked one gunnysack at a time from the roadside during his daily commute as a high school shop teacher.

We knew about Al because he had come to the local farmers market several times over the previous couple of years, talking up the farmers to see if any of them wanted to cut balsam brush from his trees to wholesale to wreathmakers. The Christmas tree groves were starting to get away from him, producing more greens than he could harvest himself, and as a consequence the paths were closing in and the Christmas trees were growing too tall. In another few years, much of the grove would be too big to sell.

None of the farmers Al talked to wanted to cut greens, and neither did we. But we moved into the farmhouse in 2008 right as the recession hit, and jobs were scarce. I found work that first winter in a shop building and repairing wood and canvas canoes, and the next summer as a zipline tour guide, while Cecilia worked as a cashier at the local co-op. We were young and poor and had a baby. Unlike so many of our peers, who were following a career from city to city, we had chosen to live in this rural town that we loved and to scrabble together a living any way we could. We also deliberately chose to have kids while we were young, before we figured things out financially, both because we wanted to be young parents since we had met early in life and had that opportunity, and also to line up the busy infant and toddler years with the beginnings of our careers rather than trying to shoehorn them into the middle.

By the second summer, we decided that I would work with Al in the trees and see if it was something I might like doing. I worked for $7 an hour, and quickly learned that it was sweaty, exhausting, and uncomfortable. Al pruned his trees with a pole pruner, holding the tool in one hand and pulling the cord that moved the blade with the other. Because the trees were so tall, we usually held the handle as close to its end as possible, with its butt braced against our forearms and the entire thing lifted over our heads in order to reach the highest tree tops. This took tremendous arm, shoulder, and pectoral strength, and at first I could only manage it for a few minutes at a time. After several years, I

The dogs playing in the snow by our side porch. Al rented the house as two units, each with an upstairs and downstairs. Our half was the portion to the left, which faced up the meadow and across the driveway to the barn.

built up the required muscles to the point where I didn't much notice it, but I was reminded how hard it was when both my brother, who is extremely fit, and my sister-in-law, a world-class Ultimate Frisbee player, each took a turn helping prune trees and lasted only a half hour.

The Basics

That first fall I helped Al cut greens, haul trees out to staging piles, and load them onto his truck for delivery. Growing Christmas trees, it turns out, is not complicated on its face. Al conveyed the basic information of how to prune trees and harvest greens and trees in about five minutes for each task, but in the years since there has been a continuous learning of nuance that I assume will persist for the rest of my life.

The five-minute version of how to prune trees, for instance, is to cut back the top leader and the four branches just below it by about a third or half, which redirects the growth hormone down the stem for that year, causing the trunk to bud out more and produce more branches, leading to a fuller tree. You also need to trim any branches in the top half that extend past the desired shape.

However, individual trees display wildly different growth characteristics, and some need to be pruned aggressively while others almost not at all. How much to prune is also affected by the demand for trees in a given year, and whether you are trying to stall production with heavy pruning (low demand) or speed it up by pruning as little as possible (high demand). Sometimes a tree has two leaders and you need to decide which should stay and which should be cut. Often, pruning is much more about reducing the number of trees and sprouts coming

One year's growth on the trees. These particular balsam tend to grow about a foot a year, but this amount varies with location and from tree to tree. These are ready to be pruned.

out of each stump so that everything is nicely staggered in size and has enough room to grow evenly. But to do this well you need to anticipate years in advance how much space things will need as they grow and how much space will be freed up by harvesting.

The five-minute version of harvesting greens (most of which are balsam from the same stumps that produce the trees) is to only cut up to half of the growth on a given stump, and to cut branches at a junction to avoid leaving a sharp stub that might poke someone in the eye. Yet how much to cut very much depends on the vigor of the stump in general (basically all branches on weak stumps should be left, while strong stumps with a young tree can be cut back quite hard). If you are carving out a path, it is important to compensate for removing all the branches on one side by leaving all the branches on the opposite side. It is also important to cut long enough branches, as these are more efficient both in the harvesting process and when breaking branches down into smaller pieces to make wreaths. Two- to 4-foot (0.6 to 1.2 m) branches with multiple stems are ideal. How you harvest has a big impact on efficiency as well. You need to choose an area with strong growth (step 1), gather stems into your hand until you can't hold any more (step 2), and then shift everything to your arm and continue cutting handfuls until your arm is stacked full of branches (step 3), before bringing the armful to the central pile. You need to be able to build a bundle that won't fall apart when you pick it up and put it down multiple times; to judge, by eye and without a scale, that a bale weighs 50 pounds (pressing down on the pile to feel its spring and thickness); and you need to be able to tie it tightly enough (without breaking the baling twine!) so that it is solid and carries easily. A loose bale feels much heavier than a tight one.

Greens also differ tremendously in quality. There is a scale insect that colonizes some trees in the grove, particularly those that are large and overgrown, that you need to be on guard against. Some trees are also simply not as green as others, and this color difference becomes more pronounced during the harvest period of November and December. Certain trees will become more yellow or bronze due to wet soil, making their greens unsuitable for harvest. However these same greens, if harvested before they change (which usually happens mid-November), will remain a perfectly suitable shade of green. So there is a larger strategy of which areas to harvest first that takes into account this change in color of certain trees; the changing needs of different areas of the grove to have the stump skirts pushed back or oversized trees removed to free up smaller, better trees; along with the need for greater efficiency earlier on in the season when there are more demands on my time (meaning areas with easy truck access get harvested first, all else being equal). I also try to maintain a reserve of unharvested greens close to the cabin down at the You-Cut grove so I can always go and harvest a little more if I need it. Some of these factors shift over the years as things change and improve and go through cycles, while some do not, and how I keep track of it all is mostly instinctual at this point.

With harvesting trees, the five-minute version involves using a thin, tall stick to judge if a tree is tall enough, then cutting it down and hauling it out to the staging area. But as with the greens, certain areas need to get cut first before the color changes, while other areas have truck access that deteriorates if the weather remains warm, and without an accurate count of the number of trees that are ready for harvest, I need to judge whether or not I can say yes to wholesale requests that come in by just following my gut sense. Occasionally a tree is perfect in every way, but most of the time there is something imperfect about it. How much imperfection is acceptable? How much is actually desirable? Cutting a tree swiftly with a handsaw is a skill in itself that requires mastering to be efficient, and in situations where I need to harvest a tree every two minutes (that includes hauling time), I need to make decisions quickly and get it on the ground fast. This means aggressive, accurate cutting. When a tree falls, its top leader can snap off if you aren't careful, particularly when the weather is below freezing, so you need to control its fall. I try to cut trees in pairs and then haul the pairs to the nearest large trail, where I leave the cut trees scattered like bread crumbs so I can take a quick tally before hauling them out to the staging area.

Staging and loading trees are also nuanced tasks. When piling up the trees, I take extra care to keep all the cut ends visible so I can accurately count how many I have without needing to move them again. If I fail to do this, I inevitably find one or two extras that got hidden at the bottom. As I stage them, it is useful to roughly sort the trees by length, since they come out of the grove in a random order but should get loaded onto the truck with the tallest trees first and the shorter ones on top. I cut trees as close to the actual delivery day as possible, as it is not good for the longevity of the tree to sit around in a pile in the sun, even if temperatures remain cool. Storing them in the shade is better but not always possible on the farm, so I generally try to cut the day of delivery or the day before. Since a full pickup load is twenty-five to thirty-five trees, it is helpful to stage trees in such a way that I can load all of the trees from a given pile at once, without leaving an orphaned tree here or there that needs to be combined with a different pile later on. Finally, it is important to leave piles uncovered, even though you might think that covering a pile with a tarp would shade it. I did that once and the microclimate created just under the layer of the tarp got hot enough to scorch the tree branches in contact with the tarp, killing the needles and ruining a bunch of trees.

The Mindset

What is the value of learning all of this if you are not and never will be a Christmas tree farmer? The value is in being aware, going into a new venture, that every part of a process is worth scrutiny. Everything matters—every tiny aspect of how something is grown or made, how it is marketed and sold, how you keep the books and pay your taxes.

Two life experiences in particular taught me the importance of paying attention to process. The first was my time working on sailing ships. Over the course of seven years, I periodically volunteered and then worked professionally on sailing ships ranging from historic replicas to oceanographic research vessels. On these ships, there are many wrong ways to do things and only one right way. Very specific habits have been passed down from one generation of sailors to the next, habits that have been honed over centuries of storms and accidents and learning from mistakes. How to handle rope, plot a course, manage sails, scrub the deck, give and repeat commands—all of these things matter. On farms, the right way to perform any given task is rarely as fixed, but the belief from my sailing years that such a way is just waiting to be discovered has fueled much of my approach.

The second life experience was working on a relatively new farm in western Massachusetts, the one that my wife and I left to move into the blue-gray house. The work culture there was very different from that of sailing ships; because the farm was both young and evolving rapidly (they had recently started a dairy), not every task had a particular prescribed way to do it yet. But the farmers pushed for and expected efficiency in all actions, so during the two years we worked there, it became ingrained in me to constantly, almost subconsciously, analyze processes to make them more streamlined. Some refinements were on a broad time scale, like what crops to plant and where for the most efficient rotation or use of greenhouse space. Others were more minute-to-minute, details of how to milk cows and wash the milking cans and apparatus. When we started, it took me an hour and a half to milk four cows and do all the cleaning up. By the middle of that first season, even as more cows freshened, I could milk twelve cows and clean up in under an hour. Every step was purposeful, but I was not rushing.

This efficiency mindset is one of the most valuable things you can bring to any business you choose to start. Often, it can be the difference between failure and success. Training yourself to scrutinize every part of a process for ways you can make it better is a habit that will serve you well throughout your life no matter what you do. Most of us do this, whether we know it or not, in some aspect of our lives. Wherever you feel the most confident, whether at your job or playing a video game or a sport, whenever you feel on top of your game it is because you have understood and mastered the details.

The details I learned about Christmas trees were all learned on the job over the course of many years, and I knew none of them that first fall working with Al. My bales of greens were not the right weight nor were they tied tightly enough. I pruned the trees back excessively. But I recognized straightaway that the nature of the work suited me. Molding a wild landscape just enough and no more was both a physical and a mental challenge. There was an endless stream of small decisions, each one not terribly consequential by itself, and to be efficient you just had to make a choice and move on: Cut this but not that; remove this; make the path go over here. I am good at snap decisions, at making a choice and then not second-guessing myself. So at its most basic level, the work of the farm was a good fit for my temperament. How we imagine a particular type of work will suit us often fails to match the reality of that job's demands. Many farmers love working with cows but get ground down by the relentlessness of needing to milk twice a day. Many vegetable farmers love the early harvest mornings but find it exhausting to endure farmers markets several times a week. Knowing yourself and what you are good at and enjoy is crucial for establishing a farm that fits who you really are.

The Transfer

At the end of that first season, Al proposed that Cecilia and I take over the tree farm, and with some hesitation, we agreed. Part of what gave us pause was that although Al was slowing down, he was nowhere near ready to stop farming himself. He was looking for someone to take over 5 of 10 ten acres, but not the easy 5. The land we started renting was one large block of trees, about two-thirds of it overgrown and impassable; Al had slowly lost the energy to keep the trails open and his customer base had dwindled. The 5 acres formed an amphitheater that faced down to the dirt road, flat in the middle but progressively steeper toward the edges. There was no truck access into the depths of this block of trees, and few paths. The only way to get a 50-pound bale

The only photo of myself and Al we ever took. I'm twenty-four in this picture, and he is seventy-eight.

Leases

We are big believers in writing things down. While we operate much of our lives on trust, handshakes, and gentleman's agreements, leasing land or a business is definitely a time to get things in writing.

- Find a lease boilerplate online similar to the arrangement you want.
- Hire a lawyer, even if you are poor, and make it a local one, not some online lawyer consulting firm. Being poor, the ramifications of losing money are greater to you, so *hire a lawyer*.
- Have a friend or family member who is not actively part of the lease (in our case it was my father) come to any meetings and take notes of everything that is agreed upon, and then copy and distribute these notes to all parties. Ask that the other party have at least one other witness to meetings as well.
- Don't assume a handshake will hold up in the face of large amounts of money, or shifting financial footing, or outside influences. Even when you think you have thought of all possible contingencies or developments, you will not have.
- Writing things down does not, however, give you the right to be inflexible. Our leases and written agreements have helped us a number of times from having the ground shift under our feet too much, but on many occasions we have agreed to change the initial agreement to accommodate Al's feelings as they evolved over the years.

of greens was (and is to this day) to swing it up onto the back of your neck and walk it down the hill. Depending on how far into the trees you were, it could take five minutes just to get down.

In addition, the lease agreement was structured so that our percentage of profits would change over time. Each year the percentages shifted more in our favor. The idea behind this arrangement (as Al described it to us) was that for the first number of years of renting the farm, we would be selling trees that Al had tended. Rather than pay a fixed amount or a fixed percentage (as we do now for most of our acreage), this agreement would fairly pay Al for the effort he had put into trees that we were selling. Because of this, our first couple of years were lean. We put a tremendous amount of work into the farm each year and then paid most of the money in rent. While we were a little bitter at times, it was ultimately a good lesson that the value of something comes from the entirety of its history, not just the parts in which you had a hand. Our farm produces a good income in part because of the effort we put in each year, but also because of the effort we have *already* put in over the years, and most importantly because of all the effort that Al put in for decades before us. It all adds up. While time passing doesn't build something by itself, sufficient effort repeated season after season certainly does.

The Start-Up

When we took over half of Al's acres, we already had a small Toyota pickup with a very rusty frame. I bought a pole pruner, a bow saw, and some loppers and pruners. We hired a friend to build us a website and design some business cards and flyers. We plotted our future and changed our newborn daughter's diapers.

Originally, we thought we should come up with our own name for the farm. We thought of our tenure as representing a bright new future for the farm, so a bright new name should naturally follow. But everything we proposed sounded ridiculous, and we eventually realized that it was silly to think that we were coming in to change (or save) anything. The farm had a whole community of customers and townsfolk who already called it the Pieropan Christmas Tree Farm, even though, interestingly, Al's official name was the Christmas Tree Farm in Ashfield. That name probably made sense in the 1950s, but in the ensuing decades, a more prominent Christmas tree farm had come along, and if you told someone that you ran the Christmas Tree Farm in Ashfield they would ask "Oh, you mean the Cranston Christmas Tree Farm?"

Not only was everyone already calling our farm the Pieropan Christmas Tree Farm, but since that wasn't the name Al used, it was actually available as far as the town and state were concerned. After

Bow saws for the
You-Cut grove get hung from
nails on trees at the edge of the road.

hemming and hawing about it for several months, we realized that it would be wise to give people a sense of continuity rather than asking them to pretend that this place they had known for years was all of a sudden something new. Honestly, everyone probably would have kept on calling it the Pieropan Christmas Tree Farm even if we *had* changed the name.

That first year was a scramble. I pushed into the trees as best I could, creating trails where there were none. I built a crude little bridge over a rivulet that Al had always let people hop over. I got the tree crews who were doing preemptive pruning around the power lines to dump free loads of wood chips up by the house, and hauled load after load in a wheelbarrow down to the grove to shore up the muddiest sections of trail. When the mud sucked it down like it was never there, I piled on more. I bought old bow saws at every tag sale I came across, and outfitted any that needed it with new blades. I made a template of our name and logo out of cardboard and used it to paint a dozen signs. These were given an appropriately pointed arrow and staked at intersections for miles in every direction to guide people toward the farm. I spent every free hour that summer out in the grove, pruning trees as the sun went down and getting soaked with dew on weekend mornings. My shoulder and arm ached from holding the pole pruner fully extended for hours on end.

When the first holiday season started, we were still struggling to find wholesale customers. Because Al was still farming half the Christmas trees, he kept all his customers; we soon found that any-one looking to buy greens or wreaths already had a source. Our only breaks came from contacts from our farming days, and it turned out that a small fraction of the bigger community-supported agriculture (CSA) farms down in the valley were interested in making our wreaths available to their customers on a preorder basis. Winter CSA shares were just becoming popular, and we thought that this would be an opportunity for us to market some wreaths, but the share pickups were scheduled to occur just before Thanksgiving and Christmas, respectively, with the former too early for people to be thinking about Christmas decorations and the latter too late because they had prob-ably already bought one. Despite these setbacks, business trickled in as we hustled, and although our lease stipulated that we pay Al the majority of the money for the trees, we could keep all the money from selling greens or wreaths. We clung to those tiny orders, for they were ours and ours alone.

The Transition

Taking over someone else's farm is a tricky business, but a scenario that is becoming more and more common as property values increase and young farmers find themselves looking for an alternative to buying land and starting their own farm. There is growing recognition that a farm is more than just the land; it's also the brand, the customer base, and the network of suppliers and wholesale relationships that have been built up over years. Taking over an existing operation comes with its own challenges, however. Time after time we were blindsided from trusting too much our gentleman's agreement that Al wanted us to take over the farm, and time after time our business was saved by that same trust. After our first year, Al wanted to change the initial agreement we had all signed, which stated that we would start by giving him 75 percent of the money and that his cut would depreciate over four years by 25 percent a year. He felt that four years was too short, too generous, and he wanted to change it to seven years, which is how long it takes for a sprout to turn into a salable tree. We met with Al and his family and, supported by his wife and grown daughters who defended our position, hammered out a new agreement where the amount we had to pay Al would decrease by 17 percent (instead of 25 percent) over four more years, for a total of five years instead of the seven he wanted. All subsequent leases would change by 20 percent for five years. We were to find that this pattern of shifting expectations would continue.

The worst moment came a few years later, when the young man leasing the house Al had built at the top of the slope of our grove of Christmas trees, a house whose lot comprised the middle third of our grove, offered to buy the house from Al. He also proposed to buy another 3-acre lot that formed the right half of our grove. Al said that was fine by him, but never told us. When we heard about the possible sale from one of Al's daughters, we immediately went to talk to the young man, with whom we were friendly, and tried to strike a deal with him identical to the one we currently had with Al. But he was reluctant to promise anything, talking vaguely about wanting to keep his options open, and left us with the very real sense that he would refuse to continue our lease.

Devastated, we arranged a meeting with Al and his family, at which I broke down into tears trying to express how betrayed I felt, like Al had yanked the rug clean out from under us. To my surprise and relief, Al's wife Mimi stood resolutely by us. How could they sell the house to someone who they knew would not honor our agreement? Did they want the farm to continue or not?

Now, Al was the kind of guy who ran his life without consulting the women in it. When Cecilia and I first began renting from him, he asked (looking at me) if we had a chain saw, which we assured him we did. He asked (looking at Cecilia) if we were good at keeping a house neat, which we assured him we were. We chose not to mention that our chain saw was at that time Cecilia's domain and that I was the housecleaner of the two. He clearly had deeply set assumptions about gender roles, and we didn't feel like rocking that boat.

But Al was also easygoing and hated to let people down. He had accepted the young man's offer because it seemed agreeable, but in the face of my distress and his wife and daughters' firm insistence

The lower horseshoe grove (so called because it is part of a horseshoe of trees planted around three sides of a field). Just one of our 10 acres of trees.

that he stand by his decision to lease the farm to us, he changed tack and agreed that of course he couldn't do that to us, since we felt so strongly. We had dodged a bullet. After that, though, we became extra careful to include my father as a secretary and witness to any important meetings we had with Al, to keep notes and make copies for all parties. We became wary.

Because Al was still working half the Christmas trees, we were running side-by-side businesses. This could get complicated during the holiday season, when cars full of people coming to cut trees started pulling down our small road. The grove we had taken over had been Al's You-Cut grove for decades, so a lot of people simply kept coming there. But Al made sure people knew he was still in business by standing down at the foot of his driveway, the access point for his remaining groves of trees, and chatting with anyone who stopped to say hello before they reached us farther down the road at our block of trees. I resigned myself to this, but got upset when I heard from several customers new to the grove that Al had pointed them down my way only when he had seen my flyer in their hand. Given that Al had not advertised in years and we had just written several press releases, handed out dozens of flyers, arranged for several fund-raisers, and built a website, we felt pretty certain that any new customers that showed up were there because of us. I did go talk to Al once or twice to make sure we were on the same page, and I also put up some large signs near where he was standing to try to direct people farther down the road. Ultimately, I just had to wait it out.

Looking back, I don't begrudge Al the choices he made; just as we were trying to figure out how to ramp up our business and make a living, he was trying to figure out how to scale back while still remaining active. Having created the farm and tended it for fifty years, he wasn't ready to let go even though he needed to start. He wanted to see us succeed, since it meant a lot to him to see his work carried forward, to see the trees still being tended instead of overgrown and abandoned; yet he wanted to keep tending his own groves of trees for as long as possible. It's amazing, really, that we had as few bad feelings as we did.

These bad feelings didn't just go one way, either. Al was an incredibly hard worker, and he was worried for years that I wasn't working hard enough, that I was slacking off and not doing what needed to be done. This mindset came from decades of being a landlord and having tenants take advantage of his amiability. He had rented out the farmhouse we were living in for thirty years, and in this time he had expensive tools borrowed and never returned, been forced to deal with a barn full of random junk abandoned by every tenant who ever left, and had even been stiffed for tens of thousands of dollars in unpaid rent. This history naturally made Al worried that we wouldn't hold up our end of the bargain. One of the things I am proudest of is that we have over the years proven ourselves trustworthy enough that this is no longer his default opinion. When I visit him now, he remarks on how hard I am working and how much it eases his mind to have me taking care of the trees.

The Story

People ask after Al many times a season. Most think I am his grandson, a belief I am always careful to correct. But whether I am related or not doesn't seem to matter. They see me as a part of a narrative that links themselves and their families to this place, to Al, and in the end to me as well. I used to think that the farm was about *our* story, how Cecilia and I rescued these trees from being abandoned. I thought it was a personal journey, just the two of us, molding this place in our image. And the work *is* personal. I do much of it alone. I make decisions by myself (sometimes to Cecilia's annoyance). I spend much of my summer plugged into my headphones listening to some podcast, shirt clinging to me with sweat, squinting against the sun to see the tops of the trees I'm pruning. I spend much of the winter by myself, making wreaths or carving spoons in the hut down by the road as I wait for the day (and the customers) to arrive or depart.

But I have slowly come to realize that this view of things is wrong, that this is actually about as far from a personal journey as I could get. The farm has never been just the actions I have taken, the choices we have made; rather, it is the result of millions of actions and choices made by me and Cecilia and by Al and by everyone who has ever come to cut a tree. It is a fabric woven from these relationships over time. I am not the hero of my own journey. I am a custodian of this small part of these people's lives. When I make a wreath, it is for them. When I clear a trail, it is for them. When I paint a sign or change the website or prune a stump, it is for them. I am bound in this fabric, supported by it, and its meaning and worth are not things I get to define. It was here before me, and my job is to carry it forward and make sure it is still here after me.

Living on the land, no matter where that land is, means both reckoning with the past and looking forward to the future, and considering yourself more of a steward than a free agent. It is a classic American failing to have a confused understanding of ownership, of its meaning and responsibilities. We are taught, overtly and subliminally, that ownership means you are the decider and should be allowed to do whatever you want. But ownership also means accepting responsibility for something. If you own a house and you fail to maintain it and let it fall into ruin, then you are affecting the lives of your neighbors, the trajectory of your neighborhood, and the fabric and tone of your community. If you own a farm and you strip its fertility, pollute the water that flows through it, and poison the soil with lead and chemicals, those ramifications affect not just you but everyone that your decisions touch, rippling outward to every part of the world. Ownership is temporary, even if we don't like to think of it as such. Our choices have real consequences for whatever comes after us.

The Deal

Before we moved into our apartment in the farmhouse, I never thought I'd one day take over a Christmas tree farm. I grew up rummaging around in the woods for my childhood Christmas trees, and was never attracted to conventional tree farms with their straight rows and endless grass to mow. More to the point, however, is that I always thought that if I ended up farming, it would be animals or vegetables. I envisioned a small CSA, or perhaps using horses to gather maple sap for sugaring. I did not imagine that one day I'd be tending a scruffy, half-wild forest of trees and holding all of these people's memories and traditions in my hands.

You hoe the row you've got, though, and what life sent my way, the shoe that fit, was this crazy little farm. Which shows both how unpredictable life is, and how important it is to be able to take what you've got and make it work for you. Sometimes you kick and struggle and you get the 20 acres of flat bottomland with a house and barn and nearby market to join. More often, though, something else comes along, and you can either seize the opportunity and see what you can make of it or keep looking. I hope that sharing our story will allow us to get at some of the underlying principles that have made our business successful.

The first principle is to start with what you have, whether an existing business or simply a set of opportunities. Want to start a goat dairy? Figure out how that's going to work financially. Got a turf farm? Make it the best turf farm you can. Got a scrubby hillside that's sort of clear and you dream of growing medicinal mushrooms? Figure out how you can start with as little risk and as low a cost as possible.

The starting place is just that: a place to begin. It doesn't need to define you, and it doesn't need to be forever. Don't be paralyzed by the idea that you need to get it right the first time around. Just take stock of what resources you have (land, skills, equipment, relationships) and what opportunities you see in the market, and make a plan.

The resources side of the equation is pretty straightforward and the one that we usually focus on when we hear about a farm. You have this land, or this desire to run a dairy, or all this milling equipment. It's easy to envision, right? It's about you and what you want to do. And we know ourselves pretty well. The opportunity side of things is trickier to see clearly, in part because it requires us to understand the wants and needs of other people. Often it requires a deep understanding of a particular community and region in a way that can only be gained by living there for years. It's always worth asking yourself why someone hasn't done what you want to do in a particular place before. That's when you realize that you are just too far away from the nearest large town, or the road gets unreliable in winter, or the town is too conservative (or liberal) to embrace what you are trying to do. If you are still house or property hunting, you can put your vision first and move on. But sometimes, like Cecilia and me, you stumble on an opportunity that doesn't match what you've been dreaming, and sometimes you already own a home and don't want to move. In these cases it becomes even more important to understand the opportunities in the marketplace.

As an example, while I have never been in a position to move on this idea, for a long time I could see that no one in our community

Baseline Skills

Everyone's starting place is different, because we each have a unique mix of skills and interests and because life throws different opportunities our way. No matter what you want to do, though, there are some skills I recommend that you have under your belt. In no particular order, here they are.

Understand bookkeeping enough to do at least the rudimentary stuff yourself. We have an accountant and a bookkeeper who help us a couple times a year, but (entirely thanks to my wife, who was an office manager for another farm) we keep our own books.

Learn how to design and build rudimentary buildings, as well as how to repair damaged buildings. You don't need to achieve mastery, but knowing how to safely use a chop saw and table saw, how to use a drill gun, and how to use tools in general is crucial. Educate yourself on what the law requires in terms of when and how to apply for permits. Develop the habit of building and patching and repairing things for yourself.

Tell your story. Like it or not, successful marketing these days is storytelling. Get a smartphone. Get an Instagram account and use it regularly. It's the easiest way to improve your photography and writing skills in small, manageable chunks, and the discipline of it

was producing pastured eggs and chickens on a scale larger than a backyard flock. The demand existed, and the right setup with movable coops at a large enough scale seemed likely to provide a good portion of a living to someone able to build most of the infrastructure for cheap. All else being equal, it is always easier to do something where the demand is already evident.

Another unfulfilled demand I can see but don't have the right setup to meet is for custom plant starts. The idea is for people to buy will give you a wealth of photos to use when building a website. Work on your writing. Study how others have communicated what you want to communicate, and then find ways to adjust their template to make it your own. While it may seem silly to do so, take the time to document your life on the land. It can be just as important as the task you are actually doing toward making it all work financially.

Educate yourself about soil health, ecosystem diversity, and landscape architecture. Pay attention to landscapes and try to understand what makes them the way they are *before* you do anything to change them. Disruption is quick and easy. Strong, beautiful ecosystems take time to develop.

Learn how to tie some knots (more on knots in appendix A). Understand how to tow a vehicle out of the mud, and keep a tow strap or chain in your car or truck.

Buy a chain saw and learn how to use it safely and maintain it in good working order. Respect its ability to maim or kill you.

Learn how to schmooze with people. Farming is seen as a solitary activity, but the truth is that farming, like all of life, is about relationships. Overcoming social awkwardness and gaining the ability to put people at ease (ask questions, be genuinely curious, give people space when appropriate) is a basic life skill that you will use with customers, employees, friends, and family alike.

whatever seeds they want to try from any company or catalog, and for me to grow those seeds for them in a heated, reliable greenhouse at a scale that justifies my time. That way local gardeners can try exactly the varieties they want but don't need to fuss around with grow lights and suffer the generally inferior-quality seedlings home setups produce in comparison with those produced by a properly run greenhouse.

One last, unmet demand I notice in my region is for seasoned kindling. Where I live, almost everyone has a woodstove, even if it's just

I never thought I'd have a Christmas tree farm, but that's what came along.
Photo by Meghan Hoagland.

to supplement the furnace. With the advent of the hydraulic splitter, people started using the splintery scraps produced by this process in lieu of kindling, but there never seems to be quite enough and starting a fire under these conditions can be frustrating. If I had a grove of pine trees at my disposal I'd be harvesting them regularly and processing them into finely split, air-dried kindling to be sold by the bushel to local families and businesses.

Notice that none of these ideas, at least where I live, can provide a living all by itself. That's not just being realistic, however; it's also the point. Farmers have always cobbled together different businesses to make a living. The trick is figuring out a seasonal flow that spreads the work across the whole year. Notice that you could do just that with these three examples: cut kindling all winter to be dried over the summer, start seedlings in late winter and sell them in late spring, start the chickens on the pasture in early spring and run them until fall, when it would be time to deliver the dried kindling to customers.

The other trick is to think of these businesses at a scale in between that of a homestead and a large commercial enterprise. What does $10,000 worth of kindling look like? A whole heck of a lot more kindling than you've ever imagined, but definitely not something you'd need to be moving around with a tractor. What does it take to produce $10,000 worth of something the first year and increase by $5,000 each year? How do you support yourself for the intervening years while that income builds to a level that can support you without an outside job?

The Reality

We rarely have the perfect piece of land; there is always some way it could be better. But perfection is not the point. Every piece of land can be made productive in some way. Usually this comes from a deep understanding of what the land already wants to do. Al once tried planting a section of field to strawberries, only to find that it was too wet. So he backed off and replanted a different section of field, and let that original bit go back to pasture. Some areas of my grove are too wet to ever grow Christmas trees, so I'm planting them to basket willow instead. Ground too rocky to use a tractor on? Figure out how to not use a tractor.

Learning to farm the land you've got will mean looking your prejudices squarely in the face and letting them go. I never thought I'd have a Christmas tree farm, but that's what came along. What matters more than your preconceived ideas of a business are two questions: Is it well suited to the land? And is it well suited to your temperament? It needs to be both of these things, because otherwise you or the land will run into a crisis of faith sooner or later. In general, if you can start small and cheap, you will get a sense for whether both of these things are true without sinking a ton of money or time into a venture, and without ruining the land in the process.

Learning to farm the land you've got will probably also mean developing a new set of skills, particularly skills outside the actual growing of whatever it is you grow. Maybe (like me) that's the skill of using your hands to make things to sell. Maybe it's cooking or otherwise using the ingredients you grow to make a more complex product. Maybe it's teaching or writing, sharing what you know. Maybe it's documenting the beauty and humor and raw emotion of the process through photography or film. Maybe it's hiring out your time doing some highly skilled task or a task requiring expensive equipment or special permitting, like mowing or slaughtering or processing hides or shearing sheep or logging. Each situation is different, being a combination of your land, your abilities and personality, the opportunities that you can seize both locally and on a wider scale, the barriers to entry of that particular business (usually in the form of regulations), the amount of time and money you can invest, and the larger timing of culture and marketplace.

While the starting place will be different for every person, it is always helpful to understand the history of your land, of your farm, and of your chosen industry. This can help inform your first steps, and it can also inspire your future moves.

CHAPTER 2

Consider the History

A man named John "Wrecka" Warger once owned the land we currently lease and more, and farmed it for the first half of the twentieth century. It was a time of cleared land; even a century after the peak of deforestation in the early 1800s, you could see clear across pastures in every direction. These pastures are now almost entirely reforested, as farm after farm changed hands or fell onto hard times, or the economics of so much pasture simply no longer made sense. There used to be more than a dozen dairy farms within a mile of the farm; not a single one remains. The neighborhood is still relatively agricultural compared with many towns, however, with a small vegetable farm on one side and a larger beef operation on the other.

In 1911, Wrecka built the farmhouse we rented from Al Pieropan; when I finally got around to mucking out the basement and hosing down the concrete floor, there was his name, scrawled into the cement next to the year 1939. Legend has it that Wrecka also built the biggest of the barns partly using lumber purloined from the sawmill in the nearest town. He would drive his wagon down to the mill to get sawdust for bedding down his animals and slip a couple boards under the sawdust each time, or so the story goes. Supposedly he also got drunk in town and did headstands on the seat of his wagon while his horse plodded its own way home. Whatever his personality, he is responsible for most of the physical objects we interacted with when we rented the farmhouse. It was his slate roof that I gingerly walked along when brushing out the chimneys; his 4-foot-long rusty crescent moon of a scythe blade I cleaned out of the room in the barn that used to house the horses; and his outhouse that I refloored and made into a garden shed. It was his pig shed that finally collapsed down the gully under a particularly heavy snow, on which, when I went to dismantle it, I discovered two layers of sheet metal roofing scraps over an initial layer of wood shingles. Wrecka is the one that hauled the old switching hut up from the train yards in town to serve as a milk room. Wrecka planted the apple tree in the meadow that held my daughters' first rope swing. I have never seen a photo of him and I know next to nothing about who he was or what else he did with his life, and yet his presence can be felt across the landscape, in buildings that remain and buildings that have tumbled down to nothing, in the curve of the driveway and in the rusted slants of barbed wire resting atop frost-heaved stone walls.

In 1955, Al Pieropan and his wife, Elise (whom everyone called Mimi), bought 25 acres of Wrecka's land, including the house and barns, for $7,500. Al was teaching agriculture at the local high school, and while the land was fenced because Wrecka had been grazing cows on it, Al wasn't interested in keeping livestock except for the occasional flock of chickens (sixty years later, when we asked Al if we could have chickens at our house, he said yes and got great satisfaction from listening to the rooster crow and seeing our hens stalk around the meadow). It was an era of working landscapes, however, and Al wanted to do something productive with the land. He found his answer for what to grow in Linwood Lesure, who operated a Christmas tree farm on the other end of town.

Linwood was a big deal. He had started experimenting with coppicing Christmas trees on his own 700-acre farm in 1936. At that time, all Christmas trees were grown as part of natural stands of trees and were a by-product of the timber industry. Linwood realized he could keep the stumps alive if he left some live branches below the cut, and he proceeded to plant out much of his land. He won awards, was the president of the National Christmas Tree Association in two different years, and in the early 1950s he taught Al what he knew.

Who knows how many other people came to visit Linwood's farm over the decades, sought out his advice for their own operation, and dreamed of one day having acres of stump-cultured trees. I would

New buds on the balsam.

wager that none of them are still around. It is a testament to Al's stubbornness and follow-through that he set out to do this thing and, unlike so many, kept with it. By the 1960s, Christmas tree farming had begun to switch to planting the rows of trees that we see everywhere today. The brief couple of decades where trees were deliberately grown as a crop but were not grown from seedlings every time was over. While it is certainly possible to start stump-culturing trees that are planted in rows, over several generations they spread and sprawl and make it impossible to reliably drive a tractor between the rows. For many getting into the business, it was more predictable, more profitable per square foot, and gave a neater appearance to grow trees from seedlings in straight rows. Gradually, the knowledge that you could even coppice a conifer was lost to the public in general, and if you were going to start growing Christmas trees, there was really only one obvious way to do it. Linwood retired, and his trees turned back into a proper forest, and Al was left as one of the only people practicing stump culture in the eastern United States. I'm aware of one other farm up in the Northeast Kingdom of Vermont that may coppice Christmas trees, and stump-culturing is slightly more common out on the West Coast, where growers have appreciated the stumps' greater tolerance to drought due to their extensive root system. But here in the East, our farm and the one in Vermont are it, as far as I can tell.

Al is not one to toot his own horn, and he has never set out to educate people that coppicing Christmas trees is possible. Every other year or so the local papers like to run stories contrasting our farm with conventional tree farms, and maybe that holds people's attention for all of five seconds, but that has done nothing to penetrate the larger narrative our culture has that you plant a Christmas tree as a seedling, cut it to the ground, pull the stump, and plant again. About six years ago, Dave Jacke, a permaculturist researching a book on coppice agro-forestry around the world, reached out to see if he could come visit the farm. He lived in the area and someone told him that he should check us out. He was amazed to learn that you could coppice conifers, had never seen it done, and knew of nobody who had written or talked of doing it. Part of my own desire, the original kernel of the idea of this book, was that if nothing else, people on a broader scale should know that this is possible. Imagine an America where farmers aren't growing greens and roots right through the winter in greenhouses, because Eliot Coleman hadn't published the results of his experiments and travels. Imagine a world where farmers aren't driving chicken tractors around pastures because Joel Salatin never wrote about it. Doing something revolutionary isn't a revolution unless you share it.

A year after buying the farm, Al started planting trees, inspired by Linwood's example. He continued planting trees for the next twenty years or more, slowly filling in area after area. Early plantings were a mix of spruces and pines, the popular Christmas tree species back in the 1950s and 1960s. As preferences shifted to balsam fir (*Abies balsamea*), Al started planting balsam, which today comprises fully 90 percent of the trees. While he bought some seedlings, the majority of trees were seedlings Al pulled from the side of the road on his commute from work. Balsam requires a certain low temperature to germinate well, and although it is not quite cold enough on the farm for strong germination, by the time Al was planting balsam he was working as a shop teacher several towns to the west. His drive took him through higher, colder regions where balsam sprouted thickly along the side of the road. He would stop and fill a gunnysack on the way home, and then go out with a shovel and plant them in his fields. Over the course of decades, this dedication turned into thousands of trees spreading over 10 acres.

When I ask Al, as I have several times over the years, what was hardest about the farm, he generally waves off the notion that any of it was difficult. He is a man naturally turned to work, and prone to understatement. If I press him, though, he will allow that in those early years when the trees were young and the land was still in transition, the brambles sometimes got the better of him. There was no multiflora rose (*Rosa multiflora*) back then, its invasion still decades in the future, but blackberry and raspberry canes can be quite unpleasant enough when they really take hold. Al won't say it, but I suspect he battled a lot of brambles in his day. Now the land is mostly settled. The skirts of branches on the stumps shade out much of the ground, although certainly not all, and the ground cover is a mix of mosses,

Anne Preston tackling a thicket of multiflora rose.

ferns, huckleberry, dogwood, and willow. There are a few areas still covered with brambles, though, and every time I push through the canes, thorns dragging at my pants, I think of Al and what it must have been like to have to cut through acres and acres of them.

When you are young and relatively poor, land that is partway through transitioning from cleared to reforested is often what you can afford. Good farmland is usually still being farmed. Well-maintained forest is similarly unlikely to be for sale. Old people often hold on to properties they can no longer maintain, and so when these properties finally do go on the market, they are neglected and overgrown, or worse, tapped out from years of tenant farmers with little incentive to invest in the land's fertility. Plants that thrive in such disturbed ecosystems are often thorny, poisonous, entangling, or just plain overwhelming, as evidenced by any pasture in New England allowed to return to forest. Brambles often overrun idle fields during this early successional stage, which can take decades. More recently they've been joined by the aforementioned multiflora rose, an ornamental species from Europe that forms huge, arching thickets of canes with aggressive thorns along every inch of stem. Multiflora rose, like brambles, possesses the devil-ish trait of walking; that is, whenever a cane arches down and touches the ground, it roots and starts a whole new plant. In this way, one plant can become five in the course of three years. When I asked Al how he dealt with multiflora rose (thinking there was going to be some optimal time to root them out, or some method involving machinery), he told me he just uses his clippers, cutting it into pieces until it is just a pile of chopped-up bits on the ground. He has a good point. Multiflora rose that is simply cut at the base and left in long pieces tends to take years to decay, since the canes hold themselves up off the ground and simply dry out; these long branches can make it difficult for the farmer to return to the base of the plant the next year to cut back the fresh growth as the plant regenerates. Instead, by cutting it into bits, more pieces touch the ground where they get wet and rot quickly, and it is easy to keep cutting the rebound shoots until the plant uses up its resources and truly dies, a process that usually takes three to four years. Al used hand clippers that he always carried in his pocket, but I prefer to set aside time each year specifically to tackle multiflora thickets, and I use long-handled loppers to give me a little distance from the thorns and more leverage when it comes to cutting thick stems. As much as I dread this task, I am always pleasantly surprised at how little time it takes to chew through an enormous thicket. Something that seemed insurmountable will take two hours, leaving me sheepish and resolved to do more to root out all the rose in the grove, of which there is still quite a bit, although each year there is less.

Coppice and Standards

When I cut back rose, I also cut back any deciduous species that I don't want to keep, a necessary task to revisit every few years for any given area. This is also a form of coppicing, as these species will usually come back from the same little stump the next year, sending up multiple shoots. I've been using this to my advantage by keeping the best, straightest ones and trimming them up to give me an endless supply of the 8-foot (2.5 m) poles I need to store and transport wreaths. Each year I go through thirty to forty poles, a rate that the local woods along the road edge cannot sustain. After doing this selective cutting for about five years, I'm starting to get a good supply of poles from the grove itself.

I've also started to select the best specimens of hardwood seedlings to grow into full-sized trees, choosing ones that have never been cut before, and therefore whose trunks have clean bases so they will remain strong as they grow old. Mostly I keep sugar maple (*Acer saccharum*), black cherry (*Prunus serotina*), red oak (*Quercus rubra*), American beech (*Fagus grandifolia*), and white ash (*Fraxinus americana*), with the occasional red maple (*Acer rubrum*) and paper birch (*Betula papyrifera*). The idea is to establish an overstory of valuable hardwoods that will start to form a partial shade over the grove, making it more pleasant to do the summer pruning. Currently there is quite a lot of squinting into the sun and strategic timing to follow the shade at the edge of the grove. To keep these trees from interfering with the balsam, I prune all the branches as high up as I can, usually 16 feet (5 m) or so. This is fairly easy for the first few years, because the saplings are flexible enough to bend them over and snip off any side branches with clippers. As they get older this flexibility lessens, and after about three years of pruning they are usually too stiff to bend down, by which time the straight, branchless section of trunk is well established.

This process of establishing a stand of full-sized trees is a common practice in coppiced forests around the world. Called *standards*, these large trees create partial shade that is beneficial to the coppiced species below. In my case, the balsam don't need the partial shade, but I do. There is another motive for creating this overstory forest, however. If I'm unable to find someone to take over the Christmas trees when I'm ready to be done (whenever that is), this forest is a backup plan. I don't think the balsam stumps will ever run out of juice as long as they are properly managed. But I do think it possible that I might not find a replacement for myself. And if that happens, I want to be able to cut down all the balsam and have a beautiful forest of mixed valuable hardwoods already established and mature. In the meantime, they will keep me cooler in the summer.

Another species that is abundant on the farm is black locust (*Robinia pseudoacacia*). Although the locust seedlings all over the farm come from two separate groves of trees that are at least 150 years old, locust is considered invasive in Massachusetts and planting it is prohibited, since its native range was originally farther to the south. Locust is naturally rot resistant (better than pressure treated) and was used extensively for fence posts back in the day.

Stump-cultured Christmas trees can be a chaotic space to take in. There is a lot going on, and there is very little geometry to guide the eye. Here, a half-grown tree edges in next to a tree ready to cut. Below them, full skirts of branches are ready to be harvested.

Black locust is also incredibly thorny, so it is not an ideal species to have growing all over the grove. However, I've taken to pruning up any I find in the same way as the other hardwoods, with the goal of having a steady supply of fence posts or poles in a few years. I've even transplanted a dozen that sprang up in a pasture into an unused section of the grove to make my own locust coppice area. Locust grows faster than most trees (except willow), so in about five years I should have fence-post-sized trees that will supply me with two or three every year indefinitely, with the number increasing as the coppice stumps mature and start to throw out more rods each time they are cut.

Site Repair

Understanding the history of your land is not about preserving the past like in some museum, unless that is your thing. Rather, it serves as a jumping-off point, learning what made sense then to think about what makes sense now. It is easy to dwell on the mistakes of past landowners (if they hadn't made any, their descendants might still be farming the land and not you), and fail to see the excellent choices that were made in siting buildings, defining pastures and woodlots, and utilizing springs. When we moved into our farmhouse, I spent the first couple of years daydreaming about all the changes I would make to the house if we ever bought it, only to realize that my first ideas were not that good, and that I had reacted to the landscape without any sense of nuance. As I observed more and lived through more seasons, my understanding of what would make the house better changed. Although we didn't end up buying the house, it was a lesson I took to heart.

The concept of *site repair* offers another way to think about our propensity to make too many changes, too fast, when encountering a new landscape. It was first put forward in the groundbreaking book *A Pattern Language*, written in the 1970s by a group of architects who were trying to describe the common themes behind the architectural success of vernacular buildings around the world, buildings designed and built based on local needs, with local materials, reflecting local traditions, and by local builders. Understanding site repair is simple. First, picture a meadow. Most people when adding to this landscape would choose to put a new building in the nicest spot in the meadow.

After all, it is such a lovely meadow. But sometimes adding to a space destroys or disrupts the very things that made it lovely and took years or even decades to develop, including plants, the way animals interact with the landscape, the way the wind blows across it, or a view. Buildings are not bad; they can vastly enrich the spaces around them if sited and built thoughtfully. But they inevitably disrupt what was there before, and so the best thing is to site them, if possible, in an area in need of repair. According to the site repair principle, the broken-down wreck of a building that is slowly moldering into the goldenrod gets removed and cleaned up to make way for building something new, and the meadow stays the lovely meadow that took so many years to evolve.

Site repair is not commonly practiced on farms, in part because there is usually more land available than there is money to clean up the old foundation, and in part because we farmers are thrifty. That jumble of old equipment on the field edge? It might provide us with a part we need someday, or we might be waiting for the price of scrap metal to go up, or we see it as some sort of crazy retirement plan to eventually sell. In the meantime, we plunk the new greenhouse or shed or barn or house, even, right in the middle of the nicest bit we've got. This is exactly the sort of short-term thinking that gets farms into trouble. What would you do differently if you were to create a five-year plan for your farm to be as successful as you hope and imagine it could be? What about a ten-year plan? Twenty-five? The best farmers make the big decisions using the history of their land to inform long-term future moves, while also reading the tea leaves in the present to keep things moving in that direction in the short term. This can mean new infrastructure using site repair, but it can also mean repairing and using the structures that already exist, whenever possible.

Cecilia and I did a lot of repurposing of spaces when we lived at the farm, and we continue to do it at the house we now own. When we needed a garden shed, I mucked out the old outhouse, put a new floor down, repaired the broken window, and built a door. When we needed a space to make wreaths and cure garlic, I emptied the little greenhouse shed of its junk, repaired the roof, and repainted it. We cleaned and organized several areas in the big barn to create space to store poles of wreaths during the wholesale season, and we dismantled another shed that had collapsed down a gully rather than just let it rot in place.

Bird nests and other wildlife in the grove are
good indications of the rich ecosystem this way of
growing trees creates. When possible I keep nests
in the trees as a surprise for the customer.

Landscape Design for Site Repair

Whenever you are adding a new structure to a landscape or renovating an existing one, here are some basic principles to keep in mind:

- Whenever possible, rehabilitate the worst piece of land with any new construction. This might mean ripping down that scary old barn, or it might mean repairing and adding to it.
- Pay attention to solar orientation and prevailing winds to create warmer and cooler microclimates in the winter and summer. The particulars of this vary from climate to climate, but local vernacular architecture usually features ingenious solutions for this, so if in doubt do some research and make field trips to good examples.
- Orient new structures so that the area formed between them becomes a useful, usable space. Often this means deliberately building structures closer together, which can in turn be used to create sunny spots out of the wind or pools of shade facing the prevailing breeze.
- Be aware of the potential for any structure to be useful around its entire edge. Design an outside that creates opportunities for storage, work, or relaxation.
- If there are trees on your property, be wary of siting buildings under their potential fall path, particularly trees like white pine (*Pinus strobus*), cottonwood (*Populus deltoides*), and poplar (*Populus grandidentata*) that are prone to snapping and falling. If they are leaning, build on the side away from the lean or remove the trees.
- Consider the workflow of your farm. Site your buildings and structures to make your operation run as efficiently as possible, within the constraints of your land. If your operation is by necessity spread out, concentrate as many operational tasks into one spot as possible, but make sure tools and supplies are close at hand for each task. Sometimes this means separate toolsheds or walls for far-flung fields. Sometimes (as with me) it means that everything rides along in the truck from location to location.

The Soil

Understanding the history of your farm is by definition understanding the state of its soil. This includes knowing what areas are appropriate for your business (not too steep for any operation that will plow up the sod, not too wet for vehicle access, not too sparse for grazing), as well as whether or not your soil is contaminated. The most common soil contaminants where I live are lead and other chemicals that were sprayed on old orchards, which can linger long after the trees are gone. But contaminants can also come from past conventional farming, or lead paint residue from an old building. If you know or even suspect that your location had an orchard or building on it, get the soil tested for any possible contaminants before you invest resources, including buying. You might discover that the land is cheap for a reason.

Sometimes contamination can come from old dumps. In New England, it was not uncommon even as recently as fifty years ago for people to dump appliances and other larger trash in the woods or down a gully. Old refrigerators, in particular, can have some nasty chemicals in them. Do some soil testing if this is a concern.

Of course, it is also worth getting the soil tested to establish a baseline so that you know where you want to go and how to get there to improve the fertility and health of your land. While most agricultural universities offer soil-testing services, these are often designed to be useful to large conventional farms and can leave smaller operations taking their best guess at what exact course of action is needed. Thankfully, there are a growing number of options for small farmers and gardeners. Of particular note is the 2012 book *The Intelligent Gardener* by Erica Reinheimer and Steve Solomon, which gave me, for the first time after years of farming and gardening professionally, the ability to analyze for myself the lab results of a soil test to determine how much of each major element (not just nitrogen, potassium, and phosphorus, but also calcium, magnesium, manganese, copper, iron, boron, and sulfur) the soil has, how much it can hold given its organic matter and cation exchange capacity, and then how to blend a customized amendment mix that can precisely address my soil's needs. I used to follow the fertility regime of the farms I worked on and leased: bed prep with ¼ inch (6 mm) of compost, liming and spreading premixed fertilizer, fish emulsion fertigation for seedlings being transplanted. In many ways, this system works just fine, if a little crudely. The problem for me came during my years of gardening after I stopped vegetable farming. Until I read *The Intelligent Gardener* I was easily swayed by the most easily implemented ideas in the other books I read. As a result, my typical fertility regime was all over the map and not rooted in the reality of my land. I made compost only from materials scrounged from the very land I was amending, thereby compounding any imbalances that existed while *not* addressing any deficits. I avoided using outside fertilizers for far too long, enamored with the idea of letting the land reestablish its own fertility under my superior management, only to panic when crops failed to do as well in their second or third year, having used up some critical element that was in short supply. I crowded plants too closely together, making them subacutely water-stressed without realizing it, and I practiced a very awkward mixture of mulching and hoeing. *The Intelligent Gardener* snapped me out of this habit of following the advice of the latest book I read.

Soil Fertility in the Trees

Because our farm is a Christmas tree farm and will never be turned back into pasture, improving soil fertility with the goal of improving the nutrient density of food is not really applicable. Instead, our goals are to do what we can to improve the color of our trees (certain areas get a little too yellow as they enter winter dormancy), and to make sure our practices promote the long-term ecological health of the landscape.

The yellow color shows up in areas that are too wet; Al always attributed it to a lack of nitrogen, assuming that the extra ground moisture leached some of the nitrogen out of the soil. In these areas, tree growth is normal and color is indistinguishable from that of other trees up until the end of November, when most of the trees (but not all!) slowly begin to turn more copper in color. Interestingly, if you harvest greens or trees before they turn, the harvested greenery remains a good green color. To test if adding nitrogen would help prevent yellowing, one year I spread fishmeal in and around one section while leaving an adjacent section unamended. I chose fishmeal because, unlike seed-meal, it would be harder for birds and other animals to eat before it had

Understanding Soil

My best advice on soil is to read *The Intelligent Gardener* by Rein-heimer and Solomon. It is not a sexy book. There are no glossy photographs. It is dense. But it will couch your understanding of soil fertility in the reality of what is actually going on.

If you are thinking, *nah, I'm unlikely to read that*, then here are some of the most important takeaways:

The quality of the soil dictates the quality of the food, and the quality of the food forms the baseline of our health. A carrot is not inter-changeable, nutritionally, from one to the next. Food has widely different nutrition depending on the soil on which it (or its ingredi-ents) was produced. This is true for everything from milk to bread to lettuce to apples. Food grown on soil that has an abundance of the nutrients our bodies need transfers more of those nutrients to us, leading to greater health. This is not the same as saying that kale is more nutritious than blueberries; instead, it means that kale grown on one soil is more nutritious than kale grown on another soil. The difference is the fertility of the soils. This difference, the nutrition of food, is the next big distinction that smaller farms can make in how they bring value to their customers. As agro-industry subverts the meaning of organic, this measure will become the most important way to define your value.

Soil that provides this nutrition is a careful mix of the different elements. This mix will be slightly different depending on the qualities of your particular soil; the point is to make your soil the best it can be, given its underlying bedrock nature. To achieve this mix, you need to test your soil to see what it contains and then add a custom mix of amendments to bring the different elements into optimal balance with one another. This balance is not the result of guesswork, but rather the result of decades of research by soil scientists. Because there are limits to how much of some elements you can safely add each year, bringing your soil into balance can take several years of testing and amendments, and it is something of a moving target. As your soil becomes more balanced and fertile, it can handle more of certain elements, and you therefore can add more. This in turn translates into better crop growth and better plant nutrition. And better taste, because food grown on fertile soil actually tastes better. The key is testing, analysis, and amendment.

Don't assume that compost is the answer to all of your soil problems. Compost varies widely in quality depending on how it was made and its components. It can have sufficient nitrogen to grow a healthy crop or it can have very little. If it was made entirely with materials from your land (this includes manure from animals that did not receive supplemental food) then it will never by itself address the fertility imbalances of your soil, since these imbalances will have been reflected in the components of the compost itself. If anything, these imbalances will become more exaggerated by the act of composting. Similarly, there is a vast difference in compost made with manures brought in (or produced yourself) depending on the feed supplements the animals received (and that thus make it into their manure), the quality of their feed overall, and whether the manure was bedded in sawdust (bad) or hay (good). Manure from the horse down the road that stands around a grazed-out paddock all day is probably not worth getting. Manure from a prized stud racehorse given select-cut hay and special supplements is invaluable, but it will usually be carefully used by the owners rather than given away or sold. Any farmer producing manure worth having is not going to give it or even sell it away. If you do not produce quality manure or compost yourself, the best bet would be to try a number of different local compost producers. My

a chance to wash into the soil and get assimilated by the microfauna. While I did observe an increase in growth (in some cases the main leaders were almost twice as long as in the control plot), that fall there was no noticeable difference in color change from the year before.

Several other people have suggested that iron deficiency might account for this yellowing instead, and I intend to run a similar current favorite is a landscaper I used to work for who windrows grass clippings and leaves in with his beef cattle.

The quality of your land, at the time you start managing it, is the result of both your location and its underlying geology and centuries of other people's decisions. Mostly these decisions were to extract resources, in the form of crops or grass or the bones and flesh of livestock, and so nutrient levels are commonly far below what they could be, even if your predecessors spread manure, lime, or fertilizers. Their additions were rarely enough to offset what was taken. Bringing your soil back to health is a process of paying the overdue balance from those who came before you. Even if your land has been lying fallow for years, don't fool yourself: It would take hundreds of years for the natural breakdown of the bedrock and the biological processes that form topsoil to return your soil to optimal health. You don't have that long. You need to take matters into your own hands.

What amendments you add to your soil will vary, both because of the results of your soil analysis, and because of what you intend to do with the land. Remineralizing pasture looks very different than remineralizing soil in a greenhouse, in both the particulars and the economics. In most cases, however, spreading amendments is something that you should budget for and prepare yourself to do on a semiregular basis. While the amounts will start to decrease as your soil approaches balance, there will always be a need to replace micronutrients and other elements removed in the form of crops, and in much of the United States rainwater leaching removes calcium and sulfur from the soil naturally.

experiment to see if they are correct. Iron sulfate is relatively inexpensive and easy to spread, and the increased ability to harvest trees and greens from these areas would pay for the amendment many times over. But when I talked to another, very experienced Christmas tree farmer about it, his opinion was that the wetness of the land was the problem, not a deficiency of some element, and that until I did

The grove is home to a rich ecosystem of forest understory plants, including this swamp pink azalea.

something to alleviate this wetness (likely in the form of digging ditches), the soil moisture would continue to impede the tree's ability to gain access to nutrients. So if the iron sulfate doesn't work, I will probably see what can be done to dig some ditches. It sounds like a lot of work, but then again, I've got my whole life to figure it out.

As for ecosystem health, Al's practice (which I continue) of piling up slash and stumps and brush and letting it all rot down in place has helped to build the soil in most areas of the farm. At times I have daydreamed about buying a chipper small enough that I could drag it around the trails and chip up the smaller branches to mulch the paths, but in the end this seems like a lot of extra work, energy use, and expense for something that will happen automatically if I am just more patient. I've also come to realize that making brush piles full of branches and trunks of different sizes creates a much richer environment for insects, and the birds and other creatures that feed on these insects. If the majority of these branches were chipped, it would certainly be tidier; but it would also be impoverished compared with the current, rich ecosystem of the grove.

Of course, in the big picture, farming is always a form of resource extraction, and my farm is no different. Year after year, trees and greens come down off the slopes, and eventually something will need to be given back. While the complex root systems of the mature stumps and all the other species of shrubs and trees around them create some incremental increase in soil fertility through breaking down soil particles and converting them, with the addition of sunlight and rain, into organic matter, which then accumulates in the topsoil from the brush piles, it may or may not be enough to offset what is taken. We will have to see. Right now the trees appear to be growing strong, with almost no disease, and with optimal growth each year. If this changes over the course of my lifetime, I will have to reevaluate what needs to be done to give back.

The decisions that Wrecka Warger and Al Pieropan made before me influence me to this day, shaping my experience and my surroundings, just as I am certain—though unclear on the particulars—that my own decisions will shape someone else's experience someday. Our actions are rocks thrown in a pond, and while the ripples of some are big enough to be obvious, those of others are only distantly felt, lapping up on some far shore long after we have turned our attention elsewhere. Farming is the act of simultaneously holding short-term goals and long-term planning in your hand, of doing today what you must so that in thirty years you will be where you want to be. This is true of life, in general, but as farming is literally concerned with the growth of things, it is more obviously so. What we do now will shape what comes after us.

The right tool for the job. Large hanging scales like this are rare. This one is heavy, so it gets set up at a loading area, usually cantilevered off the truck rack, and lives in the truck cab so it is always available.

CHAPTER 3

Taking Advantage
of Core Opportunities

No matter where you live, what your land can best produce will shape your decisions and define your core opportunities. If your acreage is primarily rolling grassland, you will probably want to graze some sort of livestock. If you have a ¼-acre lot in a city, chances are good you will grow vegetables. If you have acres of forest, you will likely be using trees in some way. In this chapter, I describe the core opportunities we walked into when taking over our farm, and in so doing, will also articulate many of the important habits of evaluating these opportunities and improving upon them. These habits are applicable in any situation, as they have more to do with mindset than with what type of land you own or what type of farm you want to start.

We walked into an existing farm—and one with a clearly defined identity—that consisted of acres and acres of Christmas trees, none of which were going away anytime soon. We knew we were not going to try to start anything else, and this clarity and focus were tremendously helpful for the first couple of years. Al Pieropan had focused on two things, trees and wholesale balsam greens, so that is where we started.

Trees

Trees are the most obvious product of a Christmas tree farm, but they also take the most time to produce. A 10-foot-high (3 m) Christmas tree takes ten years to grow from a seedling, and seven to grow from stumps like ours. Seven years is long enough that it makes less sense for me to trace the arc of an individual tree than to assess the general abundance of available trees at any given time and compare it with that of years past. On a conventional tree farm where trees are grown from seedlings and planted in rows, it might be fairly easy to walk down the rows with a clicker and count the number of trees you have available for sale that year, or the next year. In the tangled jungle that is our farm, however, this is impossible. Instead, I rely on my gut sense of whether or not tree production is keeping up with demand.

Growing coppiced trees in the way that we do is more about cutting out all the unwanted stuff ("leaning on the groves," as Al used to call it) than it is a nurturing process of supporting tender little seedlings. The work is brutal in the summer months, pruning trees and clearing paths in the heat, the flies tormenting me while I squint into the sun and wonder when I can stop. Al traditionally pruned the trees all summer, which he had off as a shop teacher. Over the years, I have found myself squeezing the pruning more and more into the beginning of September, trying to luck out with some cooler weather.

It is a common misconception that my trees aren't pruned at all. People can recognize that they aren't sheared the way most conventional trees are, but I do in fact lightly prune them each year by cutting back the top leader and the four subleaders below it, and then shaping the top third or so of the tree. If I didn't do this, the trees would quickly become too leggy to meet our current standards of what a Christmas tree should look like. Under my pruning regime, the trees have a semishaped look, more natural than topiary but thicker than a forest tree. Each year, as demand increases, I find myself pruning less and less, trying to let the trees size up as quickly as possible. Who knows—maybe in another eight years I won't be pruning at all.

With tree pruning comes trail clearing, although I often schedule separate bouts of clearing trails using loppers to tackle truly overgrown areas and a brush scythe to nip back fresh growth just sprouting up.

When I prune, however, I always carry a large pruning saw and clippers on my belt so I can cut away saplings growing up next to the stumps and cull extra branches, trees, and shoots from the stumps themselves. I consider myself done pruning an area of the grove when the only yearly maintenance it will require over the next three years will be to knock back the top leaders of the trees themselves. So each fall, I push hard into whichever area offers the greatest opportunity for return on this effort. Some areas of the grove are too overgrown, and these need to be handled later when I'm harvesting greens. Some areas are harder to get to, or don't hold on to their color as well because the soil is too wet. But each year I find the next logical spot to really lean on the grove, where if I can just uncover them, lots of good trees and productive stumps are waiting to be opened up with clear paths and cutting back brush. Once that area is done, I quickly shape the trees in the rest of the groves.

Every farming task has its own window of opportunity, sometimes dictated by the flow of the broader work, sometimes by biology. On a vegetable farm, a sowing date is dictated by the requirements of the crop itself, where the farmer needs it to fall in their field rotation, the flow of goods the farmer intends to bring to market, and the historical weather data. Because weather and climate differ from place to place, these patterns are extremely local. In the balsam groves, in our neck of New England, the pruning window is from the end of June to the middle of September. By the end of June, the fresh growth—particularly the top leader, which grows fast and soft and starts out floppy—has hardened up and straightened out enough to be cut. At the other end of the window, pruning cuts made after the middle of September won't have enough time to fully heal over and fade before the holiday season. But you can certainly prune at any time of the year, and sometimes it makes sense to knock back top leaders in the winter that failed to get pruned in the summer, just to keep trees from getting too leggy.

I started out pruning in the evenings and on the weekends since I had a number of jobs over the years, all of which precluded me from just pruning when I wanted to. It was something I squeezed into my schedule. As my family sat down to dinner, I would walk down to the grove to prune until it got too dark to see. When they were headed to the lake to go swimming on a Sunday afternoon, I would be swimming in my own sweat up in the trees. The last five years or so I have been able to set aside precisely the weeks I want in September to focus on the grove, and the work, while still hot and overwhelming at times, has gotten significantly easier. It is hard to sacrifice your work/life balance to do the work that is often required to start something up. Layering an extra project like pruning a grove of balsam on top of a full plate of work is exhausting, but if you can hold on long enough, you can get yourself to a place where your land-based work becomes the main work and begins to fall into the regularly scheduled part of the day. Just don't expect this to be the case when you're starting out.

Balsam Greens

Back in the 1950s and '60s, most Christmas tree shoppers wanted pines and spruces, so Al planted the earliest parts of the groves to these species. As balsam fir started to become the preferred Christmas tree, he switched to planting those. He continued planting balsam until he had planted out all the areas he wanted to fill with trees, which in the end accounted for 90 percent of the groves. This turned out to be a good move, because balsam is the species that wreathmakers want. A significant portion of my farm's income, and a crucial step for maintaining the coppiced stumps, is cutting balsam greens, either from the stumps or from taking down overgrown trees, to sell wholesale.

The beauty of harvesting balsam greens is that the work fits quite nicely around harvesting the trees themselves. It only takes a few days to cut the 150 to 200 trees that we sell wholesale each year, and another few to deliver. I generally block out the entire week leading up to Thanksgiving for deliveries, as we have a great number of wreaths to transport as well. For the rest of the month of November, we cut greens and tie wreaths. Since when to cut the greens is frost-dependent (after a couple of hard frosts they will hold on to their needles much longer when cut), most years we start harvesting greens by the beginning of November, which gives us several weeks to cut and sell balsam to all the local garden centers and wreathmakers. We usually take a number of big orders at the beginning of the season; these taper out as everyone places smaller secondary orders tailored to their demand. This tapering corresponds nicely to our own need to pivot to making more wreaths to sell at the grove during the heavy You-Cut weekends in the beginning of December.

As I wrote in this book's introduction, I haul my bales of greens out of the grove on my back, and while over the years I have toyed with the idea of using some sort of wheelbarrow or travois to allow me to carry a bigger load, the paths are so narrow, lumpy, twisting, and otherwise encroached upon by the trees that it has always been easiest to just hoof them out by hand. I position my truck as close as I can to a central point near where I'm harvesting, and load it up with bales before driving them all down to the tarp barn for storage out of the sun.

Al used his tractor to haul out four or five bales at a time on a small platform mounted to a three-point hitch that stuck out like a beaver tail, which allowed him to maneuver through terrain where a trailer would get stuck. He also, in his later years, began to offer balsam in smaller bundles that were easier for him to handle, weighing 30 to 40 pounds (14 to 18 kg) instead of 50 pounds (23 kg). He stacked them up against the northeast wall of his house where people came to grab them as needed, putting money in the peanut butter jar on his doorstep. I started out trying to get each bale of mine to weigh 50 pounds exactly, but quickly realized that it was much more efficient for smaller orders to simply make each bale exceed 50 pounds and not waste the time removing the extra. For large orders, I set up a hanging scale (inherited from Al—this style is extremely difficult and expensive to find at this weight range) off the truck frame as I'm loading up for a delivery, and keep a running tally of the exact weight of each bale. If my order is for 25 bales at 50 pounds each, that equals a total of 1,250 pounds (570 kg); as soon as my running tally exceeds that I am finished loading. This process allows me to cut greens without ever weighing them in the field, which speeds things up tremendously. I also try to cut greens for a big order on the same day as the delivery. This allows me to weigh and load them as soon as I cut them, rather than having to unload them and shift them under the tarp, only to haul them out and load them for delivery later. It's much easier to haul them out to the truck, weigh, load, and deliver at the end of the day. If I choose my section of grove wisely and the cutting is good, I can harvest twenty-five bales by three p.m., giving me time to deliver before it gets dark. That is a $725 day, not a bad profit margin for what is essentially a by-product.

Wreaths

While Al never made or sold many wreaths (his wife and daughters sometimes made some to sell, but it was not a big focus), that first season I recognized that it was a natural fit down at the You-Cut grove to have wreaths for sale, and so taught myself how to tie them. A wreathmaking friend of mine recommended the place where she got her supplies; when I got the catalog, I was overwhelmed by the options. Did I want crimped rings or double-railed rings or smooth rings, and how many? What gauge wire did I need and how much? What color of ribbon should I get, of what material and in what size? Did I need a bow-making machine? Cecilia and I puzzled together an order that allowed us to offer three different sizes with ribbons to match, and were pleased at how swiftly our supplies arrived in the mail. I honestly don't even remember how I learned to tie wreaths—if someone showed me or if I just figured it out. I'm certain I never watched any videos or did other online research, but that is always a good place to start if you don't have a gut sense of how to begin something.

I ended up being right in my prediction that the wreaths would be an immediate hit. That first year I rigged up an awkward display rack and sold them ridiculously cheap: $6, $10, and $16 for small, medium, and large, respectively, roughly half of what I charge now ($15, $20, and $25). Keeping my prices low to start was a good move; they reflected the quality I was then able to produce and the fact that I had no repeat demand. Every purchase had to be an impulse buy. Had I started out trying to match wreath prices in the stores, I would have made less money, not gotten nearly so good at tying wreaths as a result of low demand, and might have even concluded it wasn't worth pursuing. Instead, I was able to increase my wreath prices after two years when demand began to outstrip my ability to meet it, and then increased them again to my current levels a few years later. Today, I make about $10,000 a year on wreaths alone. They consistently constitute about 30 percent of the gross sales of the farm, and I feel grateful that I gave myself the economic incentive to make lots of them right from the start.

Each year I tie between three hundred and four hundred wreaths. Depending on their size, each one takes me between five and ten minutes to tie if the greens are already cut. For the last several years

I've tied enough wreaths now that I can do it on autopilot. The pace is fast and each motion is precise. *Photo by Meghan Hoagland.*

The workbench where I tie wreaths. The half-moon cutout allows me to reach the far corners easily. Everything is in reach and very little movement is wasted.

Tying Wreaths

You don't need to have a Christmas tree farm for it to make economic sense to tie and sell wreaths. You can buy greens from an existing Christmas tree farm and still be profitable. Around where I live, a lot of the small-time wreathmakers get permission from landowners to take down overgrown balsam or other conifers from abandoned stands of Christmas trees planted decades ago that never got cut. If you have a CSA, wreaths can make a nice addition to your share boxes at the end of the year, and an opportunity to get people signed up for the next year. If you have a farm stand, why not keep it open? If you go to markets, any market leading up to Thanksgiving is fair game, and if you mostly wholesale, you can probably find customers for that, too.

The art of tying wreaths is about controlled chaos. Too perfect, and they look lifeless. Too uneven and they look unsure of themselves, having lost their pleasing symmetry. The trick is to ensure that each bunch of greens that forms the wreath falls within certain parameters for size and length, and once that benchmark is met to avoid making them completely uniform.

I prefer crimped rings, as they are cheap and, unlike smooth rings, keep bunches from sliding around. I also use #24 green coated wire, which is strong enough to keep the wreath together without being overkill and is camouflaged well against the greenery. For my small, medium, and large wreaths, I use 6-, 12-, and 16-inch diameter rings (15, 30, and 40 cm). The rule of thumb is that the finished wreath will extend past the ring by about 6 inches on each side, for a total of a 1½-, 2-, and 2½-foot finished diameter (46, 61, and 76 cm) for the completed wreaths. While this might not sound like much, even a small change in diameter makes a big difference in how big a wreath looks. Each year, I sell about twice as many mediums as I do smalls and larges. I order all my supplies from Kelco up in Maine, and am always amazed at how swiftly orders arrive in the mail.

Before tying wreaths I like to prep a big stack of greens, preferably enough for several wreaths, since one of the biggest wastes of time is picking up and putting down clippers. A word about clippers: I used to try to sharpen my clippers, since a decent pair costs $25, and working with them day in and day out dulls them quickly. But I could never sharpen them to the same edge that they came with, and I was struggling with tendonitis, so I eventually made the decision to just buy new ones when the old ones became dull. I can buy as many clippers as I want, but I only have one left hand. I prefer medium-sized ARS or Bahco bypass clippers (which are top-quality blades but relatively inexpensive) with simple plastic or metal handles. I don't have much use for the newer clippers with rotating handles, ergonomic grips, and multiple layers of rubber coatings, which I find to be more difficult to pick up and put down repeatedly and which are overwhelming to use. I prefer leather sheaths large enough to store them unlocked with their handles open (so that the spring of the handles pressing open against the leather acts as a friction fit) and deep enough that they don't risk falling out.

Picking each branch up from the pile by the cut end, I start snipping off appropriately sized lengths, working my way methodically down toward my hand. For each size of wreath there is a length range to shoot for: small wreaths, 6–8 inches (15–20 cm); mediums, 8–10 inches (20–25 cm); larges, 10–12 inches (25–30 cm). These lengths

Each successive bundle hides the base of the previous one.

translate to different branch junctions. Your typical balsam branch is perfectly symmetrical, with smaller branches pairing off a central stem. Each of these branches, in turn, is composed of a central stem with even smaller paired branches. These branch junctions are a consistent distance apart where they meet the central stem, making them convenient measures of length for the different sized wreaths. For mediums, for example, I tend to cut at every branch junction, leaving as many pairs together as are evenly matched, to reduce the number of stems I need to handle. For large wreaths I often leave every other branch junction and for smalls I often cut each branch junction apart. These translate into pieces that are roughly 6, 8, and 12 inches long for small, medium, and large.

I take care to have the cut portion fall nicely onto the pile on my workbench so that all the branches are oriented faceup with their stems toward me. This makes it easier to pick up the right amount for bundles, since there is no need to flip or shuffle pieces. If all my pieces fall within the length parameters, all I need to do is pick up six to eight pieces (depending on the number of doubles), bunch them in my right hand, and hold them to the ring while I wrap the wire with my left. I start the wreath by twisting the wire several times around the ring, and then spiral it three times around each bundle, pulling tight

I have hired someone to cut the greens down to the appropriate size for me to tie wreaths, which has been helpful to keep my hands from becoming overstrained. I have yet to find anyone able to cut greens as fast as I can, however. Usually my assistant can just about keep pace with me tying, so the total production time becomes ten to twenty minutes per wreath, whereas since I also cut greens at double the pace of anyone I've hired, the range if I do everything myself is seven to fifteen minutes. Don't be confused by the fact that my solo times are shorter than when I have help; the times represent the total amount required for each wreath, not just my own time. These are the ranges I use when determining the relative profitability of each size wreath, each time. Different wreathmakers spiral the wire different ways, but I find it easiest to come up through the center of the wreath and wrap down around the outside. I hold the wreath up slightly in my right hand, which also holds each bundle in place, while my left hand pulls each turn tight in a straight-armed downward motion. If I wrapped the opposite way, I'd be pulling toward myself to tighten, using my biceps, a slightly more fatiguing motion, and a detail that matters if you are doing something all day, every day.

The wreath is formed as each bundle overlaps two-thirds of the previous bundle, skewed so that one-third of the width is inside the ring and two-thirds is outside. When building each bundle in my hand before wrapping, I make sure the top branches are the best quality I have, as they will be the most visible. Scruffy pieces go all the way in the back to fill out the density while remaining hidden. When I cut the greens, I make sure to cut at branch junctions so that there are no long cut stems sticking up that will be visible in the finished wreath.

The last few bundles that complete the wreath get their stems tucked underneath the branches of the first bundles, and when I can't fit any more on the ring, I flip the wreath over, snip off the extra stem length of these last few bundles, and tie a quick hanging loop with the wire that also keeps it from unraveling.

which is a helpful rubric for planning long-term shifts in what we offer and whether or not it is worth hiring help each fall. Hiring decisions will be covered in more detail in chapter 6.

Decoration

When we first started tying wreaths we didn't know how to tie bows, so we bought a bow-making machine, which is basically a jig for forming perfect loops and holding them in place while wrapping the wire. It worked, but it was cumbersome and I never learned how to use it well, because after that first year our neighbor Jane Bucklow stopped

Wreaths with the winterberry and pinecone decoration are by far the most popular with my farm's customer base, accounting for roughly half of wreaths sold. They do not take kindly to repeated handling, however, so I don't offer them for wholesale.

by and very sweetly told us that our wreaths were great but our bows were terrible. They were too rigid, soulless in their perfection. She showed us how to tie bows by hand, which I have been doing ever since, sending out a silent thank-you to Jane every time, for helping us breathe life and beautiful imperfection into our work.

The three wreath decoration variations I use are regular bows with a knot in the middle; bows with a cluster of pinecones wired into the middle instead of knots; and, for small and medium wreaths only, a bundle of red winterberries with a cluster of pinecones hiding the base of the stems, always affixed at four o'clock on the wreath. I have tried making other wreaths, including striped wreaths made by alternating bundles of balsam with pine or spruce, wreaths made entirely from spruce and pine, and wreaths featuring a mix of all species. Ultimately, however, balsam is the fastest to work with and the easiest on my hands. Likewise, I have tried using different ribbon colors over the years, different ways of presenting the berries, using single cones

Winterberry, otherwise known as American holly, grows abundantly in the wetter areas of the grove.

instead of clusters, and a number of other ideas. Over time I have settled on these variations, in these sizes, as the right mix of beauty and efficiency. I am a big believer that limited choices makes for greater satisfaction in life, and so I resist the temptation to add more dazzle to my wreaths. I want my wreaths to stand out for the timeless quality of their materials and construction, not their dazzle.

The Art of Selling

Selling, whether in person or online, requires a level of promotion that many farmers are uncomfortable with, and a level of interaction that often runs against the grain for individuals who might be inclined to their own company. The truth, however, is that farming is as much about selling as growing, unless you sell all your crop as a commodity to one buyer. The image of the farmer as taciturn or preferring their own company just doesn't fit with the successful farmers I know, all of whom are very good at marketing themselves, chatting with customers, and generally being charming and charismatic. If that seems like a high bar, remember that they are not like this all the time. As with most people, they have a limited supply of extroversion, which they recharge by being alone.

I work alone basically every day throughout the year, with the exception of the actual Christmas tree season, and even then I am alone (or with my crew) on weekdays. On weekends, I chat and laugh with customers for eight hours straight, and I come home tired from socializing, even though in the moment it is invigorating and keeps me going. I need the quiet at the beginning and end of the day to balance out the rest, but when I am on, I am *on*, and if you do direct sales, you need to figure out how to develop this skill, too. People know when you are disengaged, grumpy, impatient, or desperate for a sale. You need to cultivate an easy cheerfulness, not only because this will draw customers to you, but because this is what will keep you happiest in the long run. I don't fake my sociability, but rather recharge the well by making sure I have time alone.

Remember that selling is about serving someone else's needs. Make sure that *what* you are selling is excellent and fairly priced, and people will be drawn to you. It is much more fun to ask people

Selling in person is about attitude. Save up the best of yourself for the customer. Here, I'm taking advantage of the "ice cream window" on the You-Cut hut to joke and chat with people as they pay. *Photo by Timothy Wilcox.*

how you can help them than it is to try to convince them to buy. If you are at a farmers market, be willing to act goofy as a means of breaking the ice. Wear a Hawaiian shirt. Wear one of those fake noses with the eyeglasses and mustache. Do some boogie dancing when no one is around. If people come to your farm, make a playlist of songs to set the mood. I listen to all kinds of music when customers aren't around, but when they are, I find the absolute best kind of music is not Christmas carols (which we've all heard too many times before), but a carefully curated list of Motown and soul classics. Put on Stevie Wonder or Aretha Franklin, and watch people start to smile and sing along. Because I don't have electricity at the grove, I use my phone or an iPod plugged into a portable speaker.

The power of music to set the mood is not a new idea. Al Pieropan used to park his truck down at the You-Cut grove, open both doors, and tune the radio to classical music. The doors did a good job of projecting the sound up the hillside, and people still reminisce about how lovely it was to trudge through the trees in the snow, opera wafting up from below. Al used this same technique to listen to music while he worked in his sections of the grove in the early years of us taking over the farm, and more than once I had to go jump his truck when he accidentally drained his battery.

As you evaluate your land and your core opportunities, take a moment to think about what level of daily or weekly customer interaction would be a good fit for you. If you love schmoozing, then a farmers market, CSA, or farm store is probably a smart choice. If you would rather not make small talk with strangers, then wholesale might be a better strategy, where you could cultivate a fixed number of relationships and touch base with them as needed. For me, the intense but brief period of social interaction from Thanksgiving to Christmas is balanced out by the rest of the year, when I work alone.

Display

That first season selling wreaths, I quickly realized that hanging them without a backdrop made them difficult for customers to notice, so the next year I built a simple 6 × 6 foot (1.8 × 1.8 m) display wall out of rough-sawn boards, with legs to prop it up and cross braces that allowed me to store two poles of wreaths under the slanting backside of the wall. I attached a tarp to the top edge of the wall, which hung down the back and protected the wreaths from rain and snow. For a few years, this space was big enough to store a reasonable reserve of wreaths, but when increasing demand and production required more storage space, we created a simple hanging system in the barn and then later built our tarp barn to store up to twenty poles of wreaths, which will be discussed in detail in chapter 4. This amount of space is not needed for most of the season, but is completely full for the couple of weeks leading up to Thanksgiving, just before it all gets delivered to wholesale customers.

The wreath wall holds three small wreaths across the top, three mediums in the middle, and two larges at the bottom. I try to keep it fully stocked at all times, and I keep the dark red bows positioned to

I gather pinecones from red pine (*Pinus resinosa*) all summer (moment of truth: Really my parents collect most of them these days) and leave them on the workbench to dry out. This causes their scales to open, at which point I bag them up and store them until I can wire them together in clusters of three.

the left, the berries in the middle, and the bright red bows to the right. This might seem a bit rigid, but aesthetically I prefer the organization, and it drives people toward the middle choice of the berries and cones, which has a higher profit margin because there's no ribbon to buy. I keep a backlog of undecorated wreaths, as well as pre-wired clusters of cones and a pile of winterberry branches, so that I can replace the wreaths as they run out. I tend to tie bows as needed, although I do try to create stockpiles of tied bows for large wholesale orders to make decorating them easier.

These sorts of marketing details may seem overly finicky, but I believe they are critical to success, whether selling at a farm stand or at a market. They create a cohesive experience for the customer, make your farm stand out from the competition, and make your operation as efficient as possible. For our purposes, it is perhaps most useful to divide them into two categories: "front-end" and "back-end" details. Front-end details are what your customer sees: the displays, the choice of materials, the aesthetic, the fonts, the colors. Back-end details are everything they don't see that is strictly for you: your system of stacking boxes, the order in which you load the truck, storage, keeping the right supplies on hand, anticipating demand through good record-keeping. Quite often, something must serve a front-end and back-end function at the same time. For instance, if you have limited storage space, you might store supplies out in the open in a visually striking way. Or consider the way you handle money when making change. What you choose to wear, even. It is important to find the right balance between functionality and the public image you are trying to create. Put on a clean shirt to go to market, for example. Get a nice tablecloth. Maintain your truck.

FRONT-END DETAILS

This list isn't meant to be comprehensive but I've tried to include a few of the more important details that often get overlooked.

Elevate your market or display table. People tend to notice things better when they're at a height between their belly buttons and shoulders. Most folding tables are too short, meaning customers often walk by without actually seeing what you have to offer. Use cinder blocks to lift them up, or better yet, lengths of pipe that can slip over the existing table legs.

Use a full tablecloth. Make sure your tablecloth extends all the way to the ground to hide any bins or tubs you might have underneath. Solid-patterned tablecloths keep the customer's focus on what you are selling rather than on the cloth. If you use a bare wood surface (like the display area at my farm), use rough-sawn lumber, if possible. The imperfection of the wood makes anything you put on it look more perfect by comparison.

Hang your signs high. While it might be easy to prop a sign on the ground against the table, this makes it almost invisible, both because people don't read signs unless they are literally in front of their nose and because anyone standing in front of it will block others from reading it. Also, don't assume if you put up a sign explaining something (buying procedure, prices, etc.), that you've done enough and the problem is solved. Signs rarely get read, and understanding which signs work and which do not only comes from a long process of observation and experimentation.

Present yourself well. No one needs to see your dirty clothes. Even if you think you are portraying an authentic view of what it takes to farm, I guarantee that's not the association created in people's minds. You don't want your heirloom tomato associated with your holey T-shirt. Similarly, maintain your vehicle. It doesn't need to be pristine, but it does need to look like you make an effort to patch the rust and apply a reasonably matching spray paint. Remember, every aspect of yourself on view is part of your image in the public eye. You don't need to be squeaky clean and brand new. But you do need to present yourself as someone who maintains a certain level of cleanliness and order; remember that the quality of your product is linked in people's minds with their impression of you.

Consider every aspect of the design. When you are creating a space, whether a vegetable stand or a market booth setup, keep things as clean and uncluttered as possible. Ask a friend for their first impression. Ask yourself what is essential and what you can do without.

Don't flash your money around. When making change, either use a money box, which keeps the whole transaction out in the open, or, if you keep all the bills in your pocket or money belt, keep a small wad of ones and fives in one pocket to make change with and keep the rest in the other pocket. Nobody wants to see you pull out an enormous wad of cash just to hand them two singles. Keep that to yourself.

BACK-END DETAILS

Back-end details are about process. This means all parts of the process, not only the actual fieldwork but also sales and office work. Again, in no particular order, here are some important things to consider.

Packing is important. If you regularly need to set up and break things down (at a farmers market, for example), think about packing order. Are the things you need first in the front or on top? Are there things you don't need? Can you pack all of your small items in a designated tub?

Stuff breaks. Do you keep an inventory of spare parts so that if something breaks it doesn't stop your work? What is most likely to break? What can you do about it ahead of time to prepare?

Stay organized. Do you have a designated space to pay bills, enter sales, collect receipts, and prepare deposits? Keep an accordion file to collect receipts. Keep an adding machine, and learn how to prepare deposits the way your bank wants them. Regularly back up your computer to an external hard drive. Keep a filing cabinet or box to collect bills, statements, and tax documents.

Be efficient. Analyze your production methods to determine where you can improve either production capacity or efficiency. Low-hanging fruit includes simply doing less (in a messy, semiwild system such as my farm, it's not always clear what I need to be doing as opposed to what doesn't need to get done, so doing less is always an option worth exploring; see following section), and it also includes doing more (counterintuitive to the previous point, but sometimes scaling up production means that you make better use of the infrastructure you already have and the time commitments you have already made). Alongside these big changes, there are innumerable ways you can tweak your processes to save time and effort. Always be thinking about this.

Be neat. When you are done with a task, put things away. Have places where things live and stick to that. Having a habit of cleanliness and neatness helps with both back-end and front-end details.

The Customer

You can't have sales without customers, so when determining your core opportunities, you need to consider how to best serve your customers' needs. In some ways it is folly to think only of what you want to do and what the land can do and what strategy your location suggests without taking into account, every step of the way, whether you are producing something that has value.

Value is the combination of quality and demand. If there is high demand for what you offer but its quality is poor, then your reputation will be damaged by its low value. If you produce amazing specialty vegetables but no one at your local market wants them (low demand) then their value is also low, no matter their quality. You can control quality by your own commitment to improvement and excellence. Demand is largely a function of the identity and reputation you have created, and how well you match your marketing strategy to the reality of current market desires. Depending on your product, you might need to market directly to high-end chefs. It might be that you need to focus your efforts online rather than in a physical shop, not because you live way out in the country, but because that's where there's enough demand to justify your time. It might be you need to only sell at a bustling farmers market, because of the perishable nature of what you have to offer. Ultimately, you want to generate value in what you do through developing a reputation for quality and for fulfilling the desires of your customer base. You can adjust what these desires are by shifting your customer base, or you can adjust what you produce to meet the needs of your existing customer base. You will probably end up doing a bit of both. When what you are producing has this type of value (meaning you have a reputation for excellence and your customers really want what you have), you will find you are not so much selling your products as allowing people to buy them. And that is a very good place to be.

Pricing

Pull, to borrow a term from Ben Hartman's book *The Lean Farm*, is generated when business comes to you. Price is an important tool for achieving pull. You can employ any number of pricing strategies, but it's crucial to understand the psychology of price and to tell a deliberate story with your prices.

One way to start selling your product is with low prices, as I did with the trees and wreaths, and then increase prices to maximize the amount of demand you can handle. This is an effective strategy when you are still learning what it takes to produce quality, when you haven't yet built a reputation for pricing one way or the other, or while you are still in the process of establishing a good alignment between the desires of your customer base and what you are producing. Low prices will encourage people who don't know you or your story to take a chance with their purchase. Whenever you are new on the scene, or are young and hungry, or haven't yet mastered your thing, low prices are your toe in the door. As demand increases, so can your prices, as your reputation will have started to build and the quality of what you are offering will have improved.

In other circumstances, however, you might find it better to match the prices of your competition or even to exceed them. This can be a good strategy when you are taking a product to a new market, particularly one with more money to spend, or when you already have the chops to do something really well. Remember that price is arbitrary, representing the middle ground between the value you and your customers each see in what you are selling. There is nothing sacrosanct about a price, and customer demand can shift this value as surely as quality can. Understand the wealth of your customer base and match your price to their expectations, especially at first. Any mismatch between your price and their expectations generates a great number of assumptions about the quality of your product, about you as a person, and about whether you are identified as one of them.

Starting low can give you somewhere to grow prices, but starting high can also give you wiggle room to reduce prices. I am not a fan of putting things on sale, because the expectation that something might be on sale later can lead people to hesitate or hold off on buying today, reducing the very pull you have worked so hard to establish. I *am* a fan of reducing prices when it makes sense, particularly if you can tell the story of why you are doing so. For instance, as my skill and speed in carving certain types of wooden spoons improves, it makes sense for me to reduce their price to increase demand by making them more affordable. While this might seem like a step backward, it is a long-term strategy to generate more pull for that work, as well as to enhance my reputation through generosity. The same outcome could be accomplished with early tomatoes, for example; by reducing prices to make them more affordable to better match the buying power of a small local market, you might capture the loyalty of customers who would otherwise shop around. Communicating your reasons for lowering prices is absolutely critical, however, or else you are just throwing money away. Tell the story on social media. Put a sign out on the table saying *NEW LOW PRICE (BECAUSE FOOD SHOULD BE AFFORDABLE TO EVERYONE)* and then have conversations with your customers about it.

I tend to start my prices low and creep them upward, although I will also lower prices as part of a broader strategy. One other strategy worth mentioning is holding prices steady, which is the equivalent over time of lowering prices, as slow inflation over years reduces the value of the dollar. With Christmas trees, for example, we kept our tree prices the same for eight years, until a year ago at the time of this writing, when we increased prices significantly. We didn't know how customers would respond to this price hike, but the most common reaction was that *it was about time*. People expect prices to increase with time (and in general all your associated costs will increase), so any long-term strategy needs to assume regular price increases. Starting too high can make you feel you can't afford to increase prices, leading to stagnant growth, while increasing prices faster than your quality and reputation can generate pull will, likewise, limit demand. Communicating clearly about when people can expect an increase in prices will help get people onboard for the change, and even generate sales as people are motivated to sign up for the CSA or place a wholesale order before the price goes up.

How to decide when to raise your prices (and what prices to set initially) is very personal. One of the worst ways to go about it, in my opinion, is to base it off what similar operations are charging. Back in our vegetable farming days, Cecilia and I were once asked by a fellow farmer if we would match their lettuce price at a farmers market. We declined, sensing intuitively that such an act was a slippery slope and that the only way to stay honest was to charge what we wanted to charge. If we came to the conclusion that we should charge the same price, then that would be fine, but to match what someone is doing because they asked when our own calculations had us adopting a lower price just felt wrong.

Customers assign both positive and negative values to low and high prices, and both strategies can lead to customer loyalty or abandonment. I have people who come every year for the low price I charge for my trees. I also know people who always go to the more expensive farm down the road, because they appreciate the quality those guys bring to the table. I have had customers who left without finding a tree they liked, in part because they see my inexpensive trees as a cheap bargain and if they don't find a perfect one that they could have gotten at a different farm for twice as much then it's not a bargain. I also have customers who have changed their tradition to come to my farm after feeling ripped off at a more expensive farm. All four scenarios are valid. You set your price trying to evoke certain responses in people, and then you adjust. In the end, people bring their own price narrative to your farm. There is little point in agonizing over getting it right at the start. Better to just assume that you will adjust your price in response to the market and your own situation. One particularly funny reason I used to agonize over prices was because I couldn't change my prices without invalidating all of my flyers. Then I realized that a run of flyers cost $30, and that by not just changing my prices and eating that cost, I was holding myself back from hundreds or thousands of dollars of sales from raising prices (although increased revenue could also come from lowering prices). It was absurd! But we lose sight of the opportunity because we focus on the initial cost, or on some small waste that a change in price would create.

In an era when prices are increasingly conveyed electronically in newsletters, on social media, or on a website, you need to worry less and less about making a flyer that you will never have to change. But remember that you can always make a flyer that sends people to your website for pricing, or just pay to get new ones! If the cost of the flyers is a sticking point, you have bigger cash-flow problems than wasting a few flyers.

Administration

Both my spoon-carving business and the scything work I do get lumped in with the farm for accounting and tax purposes. This is because each business needs to be registered separately with the town and state, have its own bank account, and may require its own insurance. By having much of what I do under the official umbrella of the tree farm, our administrative tasks are more straightforward than they would be if I created a whole new business for each venture.

These administrative tasks are admittedly my Achilles' heel. After our first daughter was born, my wife went back to work for the farm we had previously left, and eventually became their office manager as their dairy and yogurt business grew. Her ability to navigate QuickBooks, whether creating invoices, receiving payments, creating deposits, or generating reports, was so far above mine that I deferred (unfairly) these tasks to her. The last few years, however, as our need for accounting has expanded beyond what was once a short holiday season to year-round work, I've been pushing to understand these tasks more thoroughly.

Some business coaches say you should outsource your weaknesses and double down on your strengths. Hire a bookkeeper or web designer or photographer or writer that can keep things moving forward so that you can do what you do best. Others fall in the opposing camp, arguing that pushing yourself to improve in the very areas you are weakest is where you will see the most surprising results. I generally subscribe to the latter camp, but I think it depends on scale. If you are farming by yourself or with a partner, you need to be able to understand every part of the process, including those that are not glamorous, that are boring or technical or just not in your wheelhouse. You need to understand tax requirements, even though you might find it advantageous to have an accountant. You need to be able to change your website, even though you might find it appropriate to use a designer. You need to know how to keep your records in a professional manner, how to prepare bank deposits the way your bank wants them, how to reconcile bank statements and prepare documents for health insurance purposes, even though you might enlist the help of a bookkeeper.

If you run a really large farm, you might delegate more, but don't fool yourself that you will ever get to a place where it makes sense to not have at least a baseline knowledge of most tasks. On every large farm I know, the owner still oversees administrative work. No matter how large of a team you have helping you (and for our tiny business we do, in fact, have an accountant, a web designer, and a bookkeeper whose services we use each year), there is no substitute for understanding a process for yourself.

What I Didn't Know About Bookkeeping (and Life) When We Started Farming

I didn't grow up thinking I would one day be self-employed. I was artistically inclined, bookish, and always assumed that I'd work some interesting job while I pursued a larger career writing. I didn't know anyone who owned their own business. I knew very few people who worked for themselves. My examples in my extended family were careers working in academia, for the state, for nonprofits or corporations. Nothing in high school or college educated me about what I might need to know to be self-employed. There were a few opportunities to learn these skills, but I never even considered taking advantage of them because I naively assumed I was not going to need them. Here is an incomplete list of things I wish I had learned in school. It is *not* meant to be a comprehensive list of all the things you may need to educate yourself about; take the time to research for yourself the regulations and laws pertaining to what you want to do.

Learn how to use some form of bookkeeping software. We use Quick-Books, but there are a number of good options. At a minimum, you will need to be able to create invoices, receive payments, prepare bank deposits, generate profit and loss statements and year-to-date reports, and reconcile bank statements. These programs are designed to handle the needs of large complex organizations, so there is *a lot* of capability and options that are more than you will need. Don't throw your hands up and walk away, however; this is how you get paid.

Research the tax requirements for your business and your state. What are the sales tax requirements? If you are selling goods online, do you know how much tax the law requires you to collect? What about in person? Food is usually exempt from sales tax but there are plenty of other agricultural products that are not. After your first year, work with your accountant (and yes, you should have an accountant, as any accountant worth their salt will save you more money than they bill you for) to establish estimated quarterly tax payments to spread the tax burden out across the year instead of needing to have all the money in April.

Research the regulatory laws for your business and state. Some forms of agriculture have relatively few, while others have quite a lot. Who can you talk to about these requirements? Which state office will you be working with, for example, if you want to sell milk? Have the laws for handling produce changed recently? Get good at making phone calls. Then get good at asking people for help.

Educate yourself on how websites work. You can choose do-it-yourself options that help you buy a domain and then host it for you as a package deal, but you should still understand the underlying structure. And if you opt for a free version, you will need to find a separate domain-buying and hosting service.

Buy yourself a filing cabinet and some hanging files. Create a logical place to keep all the tax forms, bank statements, leases, titles, and insurance forms that you will need to keep track of. Even if you do much of this online or on your computer's desktop, print out hard copies as well. Don't rely on your computer working smoothly year in and year out. Machines break. Accounts get hacked. Buy an external hard drive and back up your computer once a month.

Buy a calendar large enough to write on easily, a day planner, and a Rolodex. Write stuff down in the real world, even if it's just as a backup to the digital version. Write down everything: mailing addresses, email addresses, social media handles, phone numbers, precise details about the orders, delivery schedules, even a reminder of how you know customers socially. There is nothing worse than having incomplete information and having to ransack your email for some bit that you thought would be easy to remember but forgot.

Save for retirement. Set up a Roth IRA that gets a small amount automatically transferred into it from your bank account each month. Every couple of years, increase the amount that gets transferred. In addition, establish a passively managed mutual fund account that is as broadly diversified as you can find (a fund made up of other mutual funds is ideal) and every year toss as much money in there as you can manage, or at least 5 percent of your earnings. Keep doing this and don't touch it. The laws of compounding interest only work in your favor if you start young (meaning now, whatever age you are, start now) and leave it alone.

Build good credit. Make sure you have strong credit so you can buy land or a house or get a business loan. That means having at least two lines of credit (credit cards, car loans, student loans all count). Use the credit cards regularly and *always pay them off in full each month*. Even if you plan to never live with debt and don't plan on needing a mortgage, do this stuff anyway so you have options down the road. Life is long and plans change.

Patience

While Christmas trees and greens represent the obvious core opportunities on our farm, the land does not produce them of uniform quality on all of our 10 acres. Some areas are in good shape; we've brought many back to peak production, with each carefully tended stump growing multiple trees as efficiently as possible, and the greens cut back every few years. Other areas are completely overgrown, with full-sized trees that will require a chain saw and a great deal of effort to remove, and that will need to be replanted once that is done. Much of the farm is somewhere in between, with productive stumps intermingled with overgrown clusters of maturing trees. Each year, I try to take down some of these overgrown patches, to convert them back into productive stumps, but it is slow going, and the need for efficiency during the harvesting season means I can only do so much of it each year.

For the truly overgrown sections, I watch and wait. While I could spend a tremendous amount of time cutting these down now, it is not yet clear to me what subsequent course of action would make the most sense. Maybe I could find an economic use for the lumber. Or I might need the additional greens in a couple of years if demand outstrips what the rest of the grove can sustainably produce. There may also be some other economic value to keeping these trees that I cannot see right now. So rather than charge ahead and make big changes before I'm ready, I do nothing.

Doing nothing is one of the most important skills to develop when it comes to land. It is easy to think we know what we need to do, only to realize after a couple of years that we were wrong and should have been doing something else. Particularly with trees, which take years and even decades to grow, change for the worse can happen swiftly, obliterating what took so long to develop; so until I have a plan for an area that is not performing at its peak, I wait. This degree of patience is needed for all of the farm, for it is the task of many years to push back the multiflora rose, to open up trails that have been neglected, to fill in muddy sections of trail with wood chips, and to tackle stumps that have become overgrown. I cannot do it all in one year. Bringing this farm up to speed will be a twenty-year process, in which I am only halfway.

The same is true for any farm, existing or envisioned. Land stewardship is a long-term habit, and one whose full potential cannot be realized in a shorter time frame. Good land, healthy land, land that is fertile and productive and humming along takes years and years to get that way. If you are lucky enough to stumble upon it, or have enough money to buy such a thing, recognize that someone else made good decisions for a long time before you came along. Don't mess it up.

If you own or can only afford to buy land that has been neglected, recognize what you are walking into. Be realistic in your expectations. Be realistic about the cost, in time and money and sweat (and don't fool yourself that you can get away with time and sweat alone, because it will definitely take money in the end), that rehabilitating the land will require. It will be worth it. If the farm is near a good market, or has other valuable attributes (a good house, or a beautiful natural landscape, or excellent neighbors, or just that it's available), then go for it. But pace yourself. Don't plow up more than you can properly manage. Come up with a realistic five-year plan for liming the pasture, or clearing the thickets, or remineralizing the fields, and then stick to it.

Some people can tolerate turning everything on its head, giving it a good shake, and then living in the chaos while it slowly improves. Renovating a house while living in it is a good example of this. I am not so able to tolerate such a state of affairs, preferring instead to keep things as pleasant as possible while making incremental changes. Partly this is skepticism that my first idea will ultimately be my best one, but it's also because I believe in maintenance, in small improvements that add up over time. Whether a car, a house, or a farm, I believe that doing what you can in the moment is better than saving up for some big overhaul in the future.

Location, Location, Location

Location is one of the most important factors that defines the core opportunities of your land. This is true not only in terms of its intrinsic attributes, but in its proximity to markets. Our farm is located about a forty-minute drive from the four nearest large towns in all four directions, although because of the regional associations, no one ever visits us from the town to the west. About two-thirds of the people who come to cut trees drive between thirty minutes to an hour and a half to get here. The rest come from within town. Because buying a Christmas tree is a once-a-year event, people are willing to drive that distance as part of a tradition, when they would be unlikely to drive that far to take part in a weekly CSA pickup or farmers market. Many of the small vegetable farms in the hill towns surrounding us struggle with relatively low demand from a sparse, spread-out population. While the bigger towns in the valley have large, booming farmers markets that hill town farms can in theory join if they can wiggle in, there is also a lot of competition from bigger farms down in the valley, who grow in some of the best soil in New England. It can be hard to compete. Some of the more successful hill town farms are in essence, like us, once-a-year events, including pick-your-own blueberries, Christmas trees, and maple sugaring operations that also serve breakfast during the short tapping season. Others specialize in wholesaling one item on which they have staked their reputation: yogurt, hay, beef, milk, or apples.

Take a moment to think about what markets you can reach and how many customers would be willing to come to you, based on where your land is situated. This will obviously differ greatly region by region, since how far people will drive on a regular basis is relative. In my part of New England, for example, we hate to drive more than half an hour, but in many places in the United States half an hour is nothing. Although some regions are also already saturated with certain types of farms, usually mixed vegetable production operations, as these have

Likely Land Scenarios

Here are some typical land scenarios and what might be successful options for farming under such circumstances. Bear in mind that these are just typical situations. Often, the most successful farms are those that manage to defy expectations.

Urban. In cities or large towns, farming opportunities are likely to be limited to a tiny land base, often artificially created. Concentrate on high-value items like tomatoes, microgreens, out-of-season salad mix, and flowers, and take advantage of the population density to sell directly to customers rather than wholesaling to stores at a lower rate. Keeping bees might be possible, but you would need to understand regulations, and might need to be prepared to provide substantial supplemental sugar if blossom sources are inconsistent and limited. If you are smart and meticulous, however, you can grow a large amount on as little as ¼ acre. On the other hand, rent and other costs are higher in cities. Soil exhaustion, air pollution, and security are particular issues, but opportunities abound to partner with schools and businesses to create streams of food scraps for composting and a large direct customer base.

Suburban/Exurban. Many of the most successful farms around us are located on the edges of the large towns in the region or in the next town over. This gives them easy access to healthy farmers markets and a high enough population density to make a farm store or CSA successful. Land in these places is often quite expensive, and while this is sometimes mitigated by conservation restrictions limiting certain land to agricultural use, that can mean additional cost and difficulties building infrastructure. We live in a part of the country that has a great number of such challenges and opportunities. There are over a dozen farm stores and as many large CSAs. Every year there is a fresh crop of young farmers who have worked on these farms and want to stick around and start their own thing, so a dizzying number of small farms come and go. Over the last fifteen years, however, many of these larger CSAs have gotten big enough that they now employ the same people year after year rather than hiring apprentices or interns. While this means there are fewer people competing to start small, new farms, it also is an indication that the barrier to entry to competing in this market has grown higher. One farming couple we know (who run the Kitchen Garden, see chapter 6) started out fifteen years ago on an acre, and through smart planning, a lot of hard work, and the leveraging of financing and grants, now farm 50 acres, produce value-added products, host an annual pepper festival, and maintain a wide array of wholesale accounts. It can be inspiring to see such examples, but it can also be overwhelming if you are just starting to wrap your head around how to get there when you are starting from zero.

Rural. Rural in New England is not the same as rural in Wyoming, which is different from rural California, which is different from rural South Carolina. But there are some broad comparisons that can be made. In general, if you are outside the range of your region's willingness to drive on a weekly basis (around us that's about thirty minutes; while many people commute more than that, if they need to drive for personal reasons for more than half an hour they will seek an alternative), then you need to either bring the farm to them by attending markets or arranging a CSA pickup in a more central location, wholesale your products, or make your operation a once-a-year event. You can do all three, and it is probably wise to try a bunch of strategies because something that gives you initial traction might not be what works after a couple of years. Every successful

the lowest barrier to entry costs and the least regulation, sometimes there is still opportunity to muscle into certain markets if you are good.

Your land and its specific attributes will winnow your opportunities down slightly. If you don't already own land, think about where you want to live, what you can afford to buy, and what you dream about doing. Our farm is sloping, rocky, and overly wet in many places. Most importantly, it is already planted to a permanent mix of conifer species, and the coppicing ensures that they will stick around, so farm I know is constantly trying out new ways of selling, whether partnering with other farms to sell at their CSA, working with a new distributor that brings goods into the bigger cities, building their own farm store, or starting whole new farmers markets to carve out a space in a crowded field. Start throwing stuff at the wall and see what sticks. Then run with that, but keep trying everything else, too. People's expectations and spending habits shift and flow over time, and you need to be poised to take advantage of these changes.

The elephant in the room in this regard is online sales, and while many farms have no clear path for selling their product online, that will likely change. Right now, an online presence is usually a way to be found, for customers to learn more about you, and to provide information. But the farther away you are from a large customer base, the more advantageous you might find selling online. In this context, you are competing not so much with large corporations or with the farm down the road, as with the small ecosystem of other similar farms who are also marketing themselves online. The tricky thing is to combine this sort of online business marketing with a strategy that also embraces local wholesale opportunities and direct local sales. Keep in mind that your competition online is doing the same; it can be hard to tell how much of someone's success stems from their online presence and how much comes from knowing a bunch of floral designers in Portland. Don't put all your eggs in one basket.

basically our choice of what sort of farm to have was made simply by deciding to seize the opportunity. Al Pieropan had to read those tea leaves, however, and decide to plant in the first place. What he saw was pasture that was never going to produce much good hay, in a town that was intensely local, in a region that was not affluent. He had a career besides farming and saw Christmas trees as a good fit for his life, the land, and the broader interest of the community. Our lease is only for the groves of Christmas trees and does not include much

pasture (what we do manage is for the purpose of maintaining truck access), but there are pockets of other tree species for which I have found a use, and there are some swampy areas for which I have plans (see chapter 9). Even the wettest, scrubbiest, most difficult piece of land can be used, if you tailor your expectations to what is suited to the conditions.

Reading the existing market often requires a familiarity with the farming scene in a given region, which takes time to develop, but knowing what farms are already doing well, what has been tried, and what gaps are yet be filled is a great way to get a toehold. Often, farmers markets are not interested in another vegetable vendor, unless you bring something unusual to the mix. It used to be common for farms to start off with vegetables and then to diversify to eggs or meat or dairy, but in the Northeast, where there is now a glut of vegetable farms, it might be a better strategy to start off with the specialized product and to branch out into vegetables after a couple of years of building a reputation, a customer base, and a presence in local markets. In chapter 1, I mentioned my three favorite ideas: pastured eggs, custom-ordered seedlings, and kindling. Part of what makes these good ideas where I live is that no one is serving these needs, but it took me several years of farming to see them as low-hanging fruit, and as I noted before, just because I can see the opportunity doesn't mean I have the right land, resources, infrastructure, or temperament to pursue them. But even if the fit isn't quite right, taking the time to assess the local scene before making a move is a smart plan. Go to a bunch of farmers markets. Visit CSAs and farm stores. Go to multiple grocery stores and see what local products they carry. Talk to people about what they wish they could get locally that they currently can't. Check out the online presence of all the major players you can identify.

Here are some of the ideas I have had over the course of a decade of observing the local farming scene here in our little part of western Massachusetts. The first three are those already mentioned, but repeated here just to keep everything in one place:

Eggs. Specifically, pastured eggs from chickens that run around a meadow throughout the year and then are given a deep litter greenhouse to muck about in with supplemental light in the winter.

A consistent supply of really good eggs is tricky to find in my region. This is starting to change, but there is still plenty of room in the marketplace for someone to raise chickens at a larger scale in a way that gains a great deal of efficiency while still providing them the best possible life. The time to get a toehold would be in the fall, as the production of all the smaller backyard producers that don't use supplemental light to maintain laying will start to dry up. If you can step in and provide a reliable source of eggs as fall progresses, that's one way to establish a customer base.

Bespoke seedlings. While almost all vegetable farms sell seedlings in the spring, I don't know of a single one that coordinates with customers to grow exactly what they want. There is a large local operation that does this on a wholesale level for other farmers, but the vast pool of home gardeners is as yet untapped. This would need to be promoted in the fall, because avid gardeners order their seeds in January. If you wait until spring to try to launch a seedling operation it will be too late. Start promoting it the year before.

Quality kindling. Given that probably 50 percent of homes in my town heat with wood, I don't know why someone hasn't tried this yet. The trick, again, would be to keep the price reasonable. We are talking quality kindling for everyday use, not artisanal kindling that is too precious to use. Scale is the key here, as you would need to have systems for creating, storing, and delivering large quantities.

Community-supported maple sugaring. Creating a collective of customers who can help gather and boil down sap in exchange for a reduced rate on the resulting syrup would help create a base of loyal customers (who would want to buy the syrup they had helped to make) in a currently saturated market. You would need two operators, one to lead the sap gathering and one to direct and manage the boiling down.

Pigs for hire. Running a number of pigs where people would pay to have them fenced in over some poison ivy or other area they wanted cleared or enriched (like a garden) could be a viable business, supplementing the money from selling the pork, and creating a customer base of people who, similar to community-supported maple sugaring members above, would have some sense of ownership and want to buy the pork.

Winter crops. A farm that specialized in only selling root vegetables that had been harvested and stored at peak sweetness after a couple of frosts, and that really managed the cold storage well so they maintained that excellence all through the winter, would have no trouble selling everything they produced. In addition, they could have a number of greenhouses dedicated to winter production of spinach, lettuce, Asian greens, herbs, scallions, and chard. By selling only in the winter, you could enjoy the summer more, command higher prices, and carve out a niche in the vegetable market that right now is just starting to be explored by the larger farms.

Value-added anything. There are farms that make hot sauce, farms that make brooms, farms that sell wreaths, yogurt, and ginger syrup. Not every idea is a winner, but you can find your niche making almost anything. The trick is to recognize that making and marketing that extra thing means extra time, with technical, legal, and marketing hurdles that you will need to figure out. It can make all the difference, however, in carving out a place for yourself in the field.

Spreading Out the Workload

The ideas in the previous section are my favorites not only because they represent relatively untapped markets, but also because the work from some combination of them would dovetail nicely across the year. Ideally, as you figure out the core opportunities of your land, you will find ways to spread the workload across as much of the year as possible. On the other hand, if you work an off-farm job, having a venture that fits into one brief, intense window of time might be a better choice. I once knew a schoolteacher who every year during February break would string a whole series of sap lines around the neighborhood, and in the evenings drive around to gather the maple sap and boil down late into the night during the short sugaring season. Al Pieropan was also a schoolteacher, and found that he could prune his trees during the summers and sell the trees on the weekends between Thanksgiving and Christmas.

Sometimes you end up with land that already wants to do one thing. Sometimes you live in a region or a specific location that makes one way of marketing more practical than others. Sometimes your own personality and passion will suggest a path, and sometimes, perhaps most often, you just try things and see what works. Whatever your path, recognizing the middle of that Venn diagram, the place where your strengths and the land's strengths and your location's strengths overlap, is the key to creating a profitable base for your farm.

———————

What and how you choose to farm is a very personal decision. It is a mix of who and where you are; if you have taken the pulse of your market accurately, you can develop a strategy that will prove profitable while being a good fit for your personality and strengths. This strategy won't pan out overnight, however. The reason most businesses fail in the first five years is that it takes time, no matter the buzz and no matter the initial success, to build a reputation that is solid enough and complex enough (by which I mean it rests on relationships with lots of customers and wholesale partners) that it can support you in a real way and withstand shifts in the market or the loss of an important customer. You might think you are solid after two years, but you won't be as solid as you will be after five. When you are just starting out, the one guarantee is that things will change. One way you can be ready for this change is to keep your infrastructure minimal, flexible, and impermanent, as we will discuss in the next chapter.

CHAPTER 4

Nimble Infrastructure

Our farm is a place of impermanence. If abandoned for just four years, it would become an impenetrable thicket, requiring tremendous effort to drag back to a useful state. Much like a bonsai that needs regular care and pruning to maintain its perpetual youth, the stumps only appear to be fixed in time, but in the grand scheme of things they are always hovering on the brink of getting away from us. And yet the trees will likely continue to produce throughout my lifetime and into the future for as long as there is someone to tend them. This permanent impermanence is at the heart of our farm, and is reflected in our infrastructure.

There are plenty of permanent structures on the farm, of course, including two large barns and multiple small barns and sheds; back when we assumed we would buy the farm (more on this in the next chapter) we used some of these spaces and dreamed of mucking out the rest and making them useful. When we bought a different house instead, we had to give up the use of these spaces, and our vision for the future changed.

Nimble infrastructure is impermanent, inexpensive (compared with the permanent alternative), and easy to adapt to the changing needs of your farm. So many farms are burdened with the carcasses of solutions that no longer apply to current problems or needs: barns that are falling down; tractors rusting in the back forty; outbuildings of every kind that once served a purpose and now serve none, even as they require more and more maintenance. Whether you view your impermanent solution as a permanent state of affairs or as a place-holder while things shake out and you determine what you really need, it can help your business stay strong by keeping overhead costs low, and reduce your long-term impact on the landscape as well.

Temporary doesn't mean cheap (although it can be) and it shouldn't mean shoddy. A cord of wood covered with a shredded blue tarp drooling strings of plastic all over the ground is not a better solution than using some recycled sheet metal roofing to accomplish the same thing. Better yet, build a cheap woodshed. The shack that is never quite finished, with exposed particleboard calving off like icebergs, is never preferable to a well-thought-out, nicely finished structure. A mass of rough lumber thrown together with little skill or planning to create a covered space is a worse option than a nicely strung tarp. Design matters. Execution matters. Make sure that your infrastructure reflects the farm's identity. Don't let cheap be the defining feature.

Nimble infrastructure means choosing the right level of permanence for your situation. Maybe you don't own your land. Maybe you don't know what the farm will actually need in five years. Maybe you don't even know what will work best *this* year. Maybe the numbers just don't add up to invest in something permanent. Maybe it would be too great of a logistic hurdle. We have found, time and time again, that the impermanent solution has been the right choice for us. You might find it to be a good approach for you, too.

Start Small

When we took over the farm, I started off copying Al Pieropan's habit of parking his truck down at the You-Cut grove and just standing around outside all day on the weekends, chatting with people, making change, and helping load trees onto cars. When we started selling wreaths, I set up a simple sawhorse workbench and made a freestanding wall to display the wreaths.

Standing outside all day in November and December can be *cold*. Because I was standing still, my feet froze no matter how warmly I dressed, even if I took regular walks. That my arrangement was oriented to face the grove also meant that I was facing the prevailing wind. I couldn't wear gloves because I needed to be able to tie wreaths,

I started off working at a trestle table out in the open, which invariably meant being very cold. *Photo by Peter Reich.*

which was therefore painful work, particularly when the weather was wet. Even so, the only weather that drove me inside was freezing rain.

After several years of this self-inflicted misery, I began to daydream of a little hut, just a small thing that I could use to get out of the wind and rain. My workbench faced the middle of the grove, and I realized that across the street was a spot, nestled right at the foot of the grove itself, where it would be logical to put such a structure, as anyone coming to cut a tree from either direction would see it. The only problem was it was awfully close to the road, as the ground was too swampy to site such a hut farther back. The logical choice, therefore, was to build a hut on wheels, so it could be simply towed to a different location if needed. This did not require a very big imaginative leap, since my wife and I had built a tiny house on a trailer frame four years earlier, when we were managing the vegetable farm down the road. When we left that farm, we were able to move our tiny house to the tree farm, where we set it up across the road as a guesthouse and office.

Hoarding

While I'm as guilty as the next guy of holding on to things that I will probably never use, and while I'm just as romantic about kicking aside the pine needles out back and finding exactly the piece of something that I need, the truth is that most things aren't worth hoarding. Here are the exceptions, followed by a list of things I've saved over the years that never got used:

WHAT TO HOARD

Lumber of any kind. Figure out where you can store it in stickered stacks, or propped on end, so that it is roughly sorted by size. Every building project yields a little extra, and over the years that can be a real boon when you realize you don't need to buy a single thing to make that chicken coop or new gate.

Sheet metal. I can't stress this enough. If you ever have the chance to grab used sheet metal roofing (particularly the corrugated kind), grab it. Just stack it to the side, and use it to cover woodpiles, roof small structures, or even smother weeds in the garden. You can never have too much.

The right fasteners. Notice I didn't say all fasteners. Repairing things can be a nightmare if, over many years, you use whatever fasteners happen to be on hand, and then all of a sudden you need to keep four different driver bits on hand just to take things apart. I hoard just three: Timberlocks (a lag bolt you drive with a drill gun), Sheetrock screws (perfect for assembling almost anything that doesn't need the strength of the Timberlock), and common nails of varying sizes.

Rope, string, and twine. While it's not worth keeping just a few feet of this or that, having a wall of hanks of rope or string is a godsend, and another example of an item where there's usually some left over whenever you buy a spool for a project.

Standard drip cap, drip edge, and flashing. It is always worth saving this stuff because usually, whether you're repairing a rotted building or building a new one, you will need some flashing. The same goes with standard tar paper felt. It's a bummer having to buy a whole roll when all you need is a 10-foot length.

Cinder block, bricks, and pavers. It is always worth keeping as many of these around as you come across, for footings and hardening paths, building retaining walls, or just pinning stuff down.

WHAT NOT TO HOARD

Plumbing fixtures of any kind. The exceptions to this are specific extras and repair kits for the actual plumbing that you use in your house, greenhouse, or washroom. You might think you would eventually find a use for that cool valve, but you won't.

Random fasteners. See above. I used to have so many of these, and then I realized I wasn't helping my life by hoarding fasteners against a time when I might not be able to go buy a wood screw if I needed one.

Extra tools. When I was just starting out, I was drawn like a magnet to old tools at tag sales, but found that only a very small percentage of these ever got used. I now keep a basic toolbox in each vehicle and a small tool wall in our mudroom for the tools that see 99 percent of the use; a small overflow gets stored in the barn. I no longer allow myself to get new tools, because I know what I need for what I want to accomplish, and I already have them. A delightful set of mini wrenches from the 1930s is not going to help me assemble the swing set we got on Freecycle; for that I need a drill gun, hammer, and socket set.

My wife and I built that first house in twenty-one days in the dead of winter, teaching ourselves along the way how to frame, sheath, insulate, roof, and install windows and doors. We made plenty of mistakes, but were motivated to forge ahead because we were house-sitting and had no other place to live when that gig was up. We moved in on what was the coldest night of the year, and proceeded to live there with our two dogs for the next eight months.

We didn't know it at the time, but tiny houses were about to become cool. When we came up with the idea, we were originally going to build a tree house before a friend suggested we consider building the tree house in such a way that we could slide the whole thing on skids onto a trailer and move it if we ever wanted to leave the farm (we were at that time envisioning buying that farm), and we thought, *Why not build it on the trailer in the first place?* That summer after we built it, the first articles about Jay Schafer and Tumbleweed Tiny Homes were in the paper (my mom thoughtfully clipped them out for us) and we discovered that we were riding the wave of a movement that was about to sweep the nation. Tiny homes are now a popular phenomenon, but back then we had to describe what we were doing as a cabin or as a glorified RV trailer.

Our tiny house had a porch (complete with hammock), sliding screen door, indoor composting toilet (also known as a 5-gallon bucket with sawdust), small woodstove, small propane cooking stove with oven, an avocado green bathroom sink plumbed very crudely through the wall with irrigation fittings to give us running cold water during the summer, a loft with our bed and milk crates full of clothing, and a space under the loft for our desks and a tiny table for eating. Electricity was conventionally wired to a breaker box that could be plugged into any heavy-duty extension cord, and we set up a coil of poly pipe on the arched roof to provide us with hot water for washing dishes or showering at the end of the day.

The entire space was 8 × 16 feet (2.5 × 5 m), with an additional 4 feet (1.25 m) of porch, and we packed a lot into it. Thankfully, we were farming, and so our days were spent either in large greenhouses or outside. We built the house because the farmers we had worked for the year before had asked if we wanted to lease their vegetable business from them so they could concentrate on building a yogurt and raw milk business (which has since grown into one of the best-known yogurt businesses in Massachusetts, Sidehill Farm Yogurt). The season before that, when we had started working for them, we had lived in an unheated, mouse-infested RV, so we went into that winter knowing that we needed a better plan if we wanted to be living on the farm in March to start up the greenhouses for early tomatoes.

Our only experience designing and building things had been a few brief stints building large furniture out of plywood for a couple of artists' studios in Boston and Cape Cod, so in preparation we got out every book the local library had on building small sheds and barns, and I began to familiarize myself with how to frame, sheath, and trim out a structure. I figured out the details of exactly how long each piece of wood should be and how it fit in with every other piece of wood by drawing detailed plans on graph paper, and then thinking through every joint and connection, and what butted up against what. I know now that there is software that will do this for you and spit out a cut list for the lumber, but I honestly wouldn't use it even if I had to do it

Moving the cabin to its summer home. We built the cabin down near the greenhouses, where it was easier to get materials during the winter, and then moved it farther out into the fields for the growing season.

all over again, because the painstaking process of thinking everything through was my education, the equivalent of doing long division by hand instead of using a calculator.

We managed to build our house in twenty-one days by making an ambitious to-do list for each day, and then exceeding what we had planned every single day. Every night I dreamed about the next day's task, roofing, and framing while I slept. While we by no means became experienced carpenters by the end of this process, it did give us confidence to tackle any task we might face on a farm or later on in our houses; if you've never designed or built anything, I recommend building something, anything, if only so you know that it's possible. Build a woodshed, or a chicken coop, or a tree house for your kid (or for you). YouTube now makes it possible to learn so much so easily that you shouldn't let a fear of not knowing how to use power tools stop you. Do some research, find a good series of videos that will teach you safe habits and basic construction techniques, and then just build something. Decent power tools are shockingly affordable right now (our portable table saw cost $150, new), and lumber for a small project is cheap. Invest some time and a couple hundred dollars in an education for yourself, and the woodshed you'll get at the end of the process will be a bonus.

Design/Build

Building the hut for the You-Cut grove was a nice chance to take all the things I had learned in designing and building the cabin and make a space that was a little more specific in its goals. Instead of needing to be a bedroom/kitchen/office/bathroom/living room/closet, the hut simply needed to be a work space, with a counter and a woodstove. It was also smaller, at just 7 × 12 feet (2 × 3.5 m), and substantially shorter, too. While it might not sound like much of a difference, those extra few feet make a big difference in how small the space feels, and a tremendous difference in the cost. We built the cabin for around $8,000, while the hut cost maybe $4,500.

I started off by scouring Craigslist for a trailer that I could use as the base. I figured there was no point in designing anything beyond some rough sketches until I knew the dimensions of the trailer. Finding the right trailer to build on takes time. You can now buy brand-new trailers designed specifically to build a tiny house on, and if you intend to move your structure more than just a few times, I would recommend you do that, as they will last the lifetime of the structure with fewer hassles and come with exactly the framing you need to safely attach the building to the trailer. On the other hand, if like us you envision moving the building rarely or never, buying a used trailer might be the way to go. If this is you, be prepared to be picky. A lot of available trailers are too rusty, too big, too overpriced, or have side rails that would need to be cut off, something I was eager to avoid. Interestingly, the trailers we ended up using for the cabin and the hut both turned out to be RV frames that had been stripped of the RV and turned into equipment-hauling trailers. The hut's trailer was cheap, seemed in good shape, and was just the size I was looking for. I hired a friend who moves strange or heavy objects for a living to come load it onto a flatbed (that way we didn't need to register it or make it road-ready) and cart it back to the farm. His truck had a long boom arm that made short work of plucking the trailer out of the snowbank it was in and dropping it down onto the truck bed.

Once I had a trailer, I could design the hut more accurately. As with the cabin, I did this by drawing the design I wanted, then transferring it to graph paper to get a sense of what different proportions and roof pitches looked like. When things were where I wanted them, I began to analyze the stud framing, starting at the floor and working my way up, drawing out each connection so that I knew exactly how the pieces would fit together and how long they needed to be. I drew stud diagrams for each wall, the floor, and the roof. One of the unusual things I chose to do with this building was to run the rafters down the length, rather than the width of the trailer, under the roof, which is the opposite of what is commonly done. I was able to do this by beefing them up from two-by-fours to two-by-sixes; fewer, longer rafters allowed me to simplify the roofing system, eliminating purlins and soffits and giving the minimal overhang a clean, low profile. I also designed the roof framing this way to make it easier to build solo.

The trailer frame was purchased just before Christmas, and I built the hut in January and February. One of the nice things about building it on wheels was that I was able to set it up in the driveway, first leveling the trailer frame on blocks so that we could easily use a spirit level to determine plumb walls. Building at the house was much more convenient than down the road at the location it would end up, as I was able to set up a table saw in the open barn door. Once a friend had helped me to frame the walls (thank you, Craig!), I was able to get the roof on quickly enough to avoid the next snowfall. I chose to sheath the walls with diagonal boards rather than plywood, as I intended to keep the wall bays open and only insulate the floor. We used a door salvaged from my parents' house, windows from a free pile down the road, roofing stripped from a collapsed shed behind the farmhouse, and a small woodstove that had been tucked away in my in-laws' shed for decades just waiting for a project like this.

Keep It Simple

The interior of the hut was, in retrospect, overdesigned. Most aspects work just fine, but there are many small things that I would do differently if starting over. Most of these changes would have made the space simpler. At the time that I built the hut, it was difficult to envision the scale that I'm operating at now. I designed the hut to handle smaller bales of greens than I now harvest, for example. I also didn't anticipate wanting to store as many supplies under the workbench as I do, and I didn't think about creating an obvious, simple counter for customers to interact with. Here are my recommendations for anyone designing a similar, small working or selling space:

- Keep the space as open and simple as you can. Don't get clever with extra storage until you've lived with it for a while and have a better sense of what you actually need.
- Think of your workflow. How do you need to interact with the space? How does the customer interact with it? What are the subconscious cues you could give the customer that would shape how they act? In my hut, most people need a place to rest their forearms, write a check, or put down their gloves.
- Imagine how your space could function if you suddenly doubled the amount of work you were trying to produce in it, and if you doubled the number of people in it at one time. Probably this means making it even simpler.

- Are windows placed high enough so that you can see outside without stooping? (This is one mistake I made; I assumed I would be sitting on a stool and designed with that height in mind, but it turns out I spend almost all my time standing at the workbench.) Do your windows command views of the important information (like whether someone just pulled up)? Do they let in enough light for your location and the time of year the space gets used?

Light

Lack of exposure to sunlight was the bane of my hut for years. Where it's located at the bottom of the You-Cut grove, the sun goes behind a hill around two thirty in the afternoon in November, and by three or three thirty it used to be too dark to see inside, even though it was still reasonably light outside. With only kerosene lamps as a light source, it was impossible to tell if the balsam we were using to tie wreaths was a good color or not, and so we had to stop work for the day. After three years of this, I finally replaced the sheet metal roof with a corrugated plastic one, cutting away the roof sheathing between the middle sections of rafters to create one giant skylight that dramatically pushes back the hour when it becomes too dark to work. Now it is as bright inside as out, with the result that we gained an hour of productivity each day, which for such a short season is a gold mine. The new roof cost $160, the value of which was paid for with the extra wreaths we were able to make on the first two days it was in place.

Heat

The woodstove is a big part of what makes the space valuable, as it would not be nearly as pleasant if it were unheated. Having a fire going on a cold day is good for my spirits and good for business, too. Customers appreciate being able to come inside and thaw out a bit. Old ladies can sit by the fire while their families tromp around getting a tree, and kids dry out after they fall down in the ditch and get soaked. Even though the woodstove we have is small, the space is so tiny that it is easy to overheat, and so the door is often propped open to one degree or another. I have gotten better at building a small fire and knowing when to feed it to avoid roasting, but it still happens. The

Even with a white ceiling and windows on three walls, the interior of the hut was too dark.

The new roof makes it as bright inside the hut as outside.

trick is to use only two logs at a time, to cut them shorter than the stove can hold, and to split them fine enough to not accumulate too much heat when fully burning but not so small that they burn fast.

The woodshed for the hut is made out of five pallets and two pieces of corrugated sheet metal. Two of the pallets form the floor, two form the end walls, and the fifth is rotated to "portrait" orientation (did you ever notice that pallets are not square?) and positioned in the center to brace the middle of the roof. The sheet metal roof is nailed into the pallets, and because the middle pallet is oriented to be taller than the side walls, the roof is curved slightly, which makes it much stronger than if it was flat. The bottom edges of the pallets are tied together with plastic baling twine. This type of woodshed holds about one cord of wood, and because it has two bays, it is easy to empty one side while loading the other. I use about half a cord each year to heat the hut, so each spring (or winter depending on the snow load) I fell a couple of the worst deciduous trees along the edge of the grove. I don't need to burn good-quality wood, so I take the trees that are struggling, or poorly formed, or just in the way. Half a cord is such a tiny quantity of wood that I should be able to harvest this indefinitely while nurturing an improved stand of larger trees that never get cut.

Storage

Just behind the You-Cut hut is our tarp barn. This structure was the answer to our storage problem when we moved off the farm and no longer had access to the barn where we used to hang poles of wreaths. At first we thought about possibly renting space in a barn down the road, but in the end that just seemed like it would mean a lot of hauling things back and forth. Then we toyed with the idea of building some sort of pole barn with rough-sawn lumber, but that sounded expensive and the area we wanted to build on is so wet we worried it would rot almost immediately. It also felt like overkill for storage that we really only needed for six weeks out of the year. In the end, we settled on a giant tarp.

The idea of a tarp comes naturally to me. For many years I worked on traditional sailing ships, where using large tarps to provide shade or cover hatches is common, so the idea of stretching a large piece of canvas with ropes to create shelter is totally normal. My time on ships also means that I have the skill and tools to repair the tarp when it gets hit by a foot of wet snow and rips a seam or a grommet tears out.

The area where it naturally made sense to rig the tarp contained a number of medium-sized birch trees that I intended to use as living posts. I harvested the three or four that weren't in the best positions and bolted them to the living trees to form a framework that the tarp would stretch across. This proved to be a mistake; since birchbark is waterproof, it kept all the sap in and the logs rotted after just one year. I replaced them all with ash and hickory poles the second year, and took the opportunity to change the orientation of the main ridgepole so that the tarp would do a better job of shading the wreaths and greens from the lower winter sun.

The ease with which I was able to swap out the rotten logs for fresh ones and the flexibility it gave me to improve the tarp barn design is one of the main advantages of impermanent structures. Instead of building an expensive, permanent storage building that would cost thousands of dollars, increase my insurance, require regular maintenance, and almost certainly not be what I need in ten years, the tarp barn cost me $300, took a couple of hours to construct, and requires an hour to rig and down-rig each year. I can change it as needed, and it asks very little of me while providing exactly what I need: a shady, dry place to store wreaths and greens for just a brief window each season.

Know Your Knots

One of the most important skills to master for using tarps (and lots of other farm tasks) is knotwork. Knowledge of knots and how to use rope in general has become almost completely pushed aside by clever, expensive solutions from the manufacturers of straps, snaps, bungees, buckles, giant twist ties, ratchets, zippers, and Velcro. There are few things so cheap and so ingenious as rope, and nothing can match all the things it can do. You can tie a knot so tightly that you need to cut it away with a knife. You can also tie a knot that you can always untie, no matter how much strain it comes under. You can tie knots that allow you to exert tremendous mechanical advantage, and you can tie knots that hold the tension created by this and never slip. You can tie knots that you can untie just by pulling on one loose end, and you can tie a knot that appears to have grown fully formed in the rope itself. There are an endless number of knots in the world, many of them associated with specific livelihoods.

Working on traditional sailing ships before I was a farmer taught me to tie knots like a second language. I taught myself to tie them in the dark, behind my back, backward and ambidextrously, with one hand aloft and hanging on for dear life as the ship bucked up and down and I fought down waves of nausea. Tying knots needed to be like breathing, something I just did without thinking, because sailing ships are operated with miles and miles of rope. Knots are the punctuation to their language.

Within the sailing world, there are thousands of knots, and some people get great pleasure from using the perfect knot for each occasion. Not me. I know dozens of knots but tie only six on a regular basis (see appendix A for instructions). These six, used in combination with one another, allow me to handle any situation. Basically, you need a way to tie a strong loop that you can untie no matter how much strain it comes under (bowline); a way to tie a similar loop without using the end of the rope but just the middle (bowline on a bight); a way to tie rope back to itself so that you can hold tension without slipping (rolling hitch); a reliable way to join two ropes together that won't come loose even if the rope goes tight and then slack and then tight again (zeppelin bend); a way to quickly tie together two ends that are under tension (square knot); and a way to quickly tie one end under tension (slipped half-hitch). The key to tying any knot is muscle memory. Practice tying them left-handed and right-handed. Practice tying them behind your back and with your eyes closed. Keep a piece of string in your pocket and practice whenever you have a free moment.

Why Rope?

I'm obsessed with rope and knots because they are versatile. I keep a couple of lengths of rope and baling twine in the truck that can handle almost any situation I find myself in, from tying down a bunch of free pallets to towing a car out of a ditch. The rope I use is potwarp that I scavenged from the beach on Cape Cod and painstakingly untangled. Potwarp is nice because it is incredibly strong, rot resistant, and, perhaps most importantly, doesn't stretch. We used to use climbing rope that someone gave us, but climbing rope is designed to stretch, making it difficult to tie something down tightly. Some other nice things about potwarp are that when cut, it doesn't fall apart the way nylon rope does, and it doesn't photodegrade the way polypropelene rope does. I whip the ends of my rope, a fancy technique I learned on sailing ships, but you could wrap the ends in electrical tape and then melt the cut ends with a lighter and it would work almost as well. If you want to learn to whip rope ends, splice, or any of the other skills I use to maintain rope and canvas, *The Arts of the Sailor* by Hervey Garrett Smith is definitely the book to buy.

Build It for Free

Nimble infrastructure also means building with whatever materials are at hand, and creating structures that are flexible in their purpose. Two such structures are critical for my spoon-carving operation (more on this in chapter 7): our woodshed and our greenhouse.

Woodshed

The woodshed was built as a more permanent replacement of our old system of stacking firewood on pallets and covering it with sheet metal. I originally planned to buy rough-sawn lumber for the structure, build a deck on concrete footings, and frame it out conventionally. But we had just bought our house and had no money for anything that wasn't a new heating system, so I figured I would see if I could build one for free. For footings, I buried about 2 feet (60 cm) of black locust trunk to support each corner. The posts are cedar and beech logs harvested from our land, the top plates are random lumber we had kicking around, the rafters are maple poles cut from just up the slope that are about 3 inches (7.5 cm) in diameter, and I had just enough old two-by-fours for purlins. The roofing is that same sheet metal that covered the firewood, and instead of a deck I spread some wood chips and laid down rows of logs to lift the bottom course of firewood off the ground. The entire structure cost a grand total of $20 for the self-tapping roofing screws with neoprene washers. One important trick when recycling old sheet roofing is to flip it upside down so all the flanges created by pounding nails through actually help to funnel water around the holes. Unless a raindrop falls directly on the hole, it will get diverted around it.

I designed the woodshed with a large overhang, which is ideal for me to work under during the warm months when I'm carving spoons. It keeps me dry in all but the most torrential downpour, is pleasantly shady, and I don't need to clean up the wood chips and shavings I make; instead, I just rake them out occasionally to keep from having too much of a mountain.

A Movable Hoophouse

The creative constraint of being poor was something we faced years ago when we moved off the vegetable farm and immediately went into greenhouse withdrawal, missing the ability to grow crops in the colder months and being able to reliably grow tomatoes and other plants that appreciate heat and being strung up. By the second spring, we couldn't take it anymore and decided to build a greenhouse, and since we couldn't afford $4,000 for a metal frame, I decided to see if I could build the frame out of saplings. I had seen boatbuilding greenhouses whose frames were made out of two layers of battens spaced out with blocks, but while these were beautiful and seemed plenty strong, I would still need to buy lumber and a table saw, both of which seemed like insurmountable obstacles at the time. I had a hunch, however, that I could use saplings to build a successful frame, even if I couldn't find any examples online of anyone else doing so.

We had a friend with the perfect patch of woods for foraging saplings, mature forest with tall, thin birch and maple saplings struggling their way up to the canopy in branchless perfection, and I harvested about twenty to build my frame. In the end I used birch for the hoops since they had fewer bumps, and maple for the braces.

That first greenhouse was far too large for our needs. But we were still used to giant farm-scale greenhouses, and so even with a 12 × 20 foot (3.5 × 6 m) footprint and 10 feet (3 m) of height it felt small. I had just discovered Eliot Coleman's experiments with movable greenhouses, so I painstakingly built this behemoth on wooden rails staked into the meadow where we were expanding our garden in a fever of farming withdrawal. The idea was that we could slide the greenhouse down the rails in the fall to cover earlier seeded crops that had grown to size in the late summer and just needed protecting from winter wind and snow.

I was pleasantly surprised how even the hoops appeared with the plastic on,
despite each of the saplings having a mind of its own.

This leapfrogging of a greenhouse to crops that grow on their own
before getting covered is a brilliant innovation of Coleman's, described
in detail in his books *Four-Season Harvest* and *The Winter Harvest
Handbook*. The problem with our setup was that the greenhouse
was too big to move without help from several strong friends, and it
only took a few years of frost heaves to mess up the alignment of the
wooden rails. The third time we moved it, we had six friends who
happened to be over for a party pick it up entirely and carry it down to
a different section of garden.

After three years, the sapling hoops were still sound but the two-
by-four rails had rotted from contact with the ground and the used
greenhouse plastic with which we had originally covered the structure
was getting scratched to the point that tomatoes were struggling to
fully ripen. I dismantled the frame, saving the hoops, and rebuilt a
smaller, stronger, tighter greenhouse. This one was just 10 × 12
feet, and a couple of feet lower as well. The bracing system for the

The big hoophouse was too unwieldy to age well or move easily. The last time we moved it, we just picked it up with a bunch of friends and shuffled it down into place. *Photo by Cecilia Van Driesche.*

hoops was a much more rigid X system on each side; the hoops were simply bolted into the sill with Timberlock drivable lag bolts that can be deployed with a drill gun with no pilot hole, my all-time favorite fastener for every situation. Four short stakes, similarly bolted to the sill, kept the hoophouse from blowing away.

The frames were originally lashed together with baling twine, but this has photodegraded over the years, and I now replace worn-out lashings with regular clothesline, which so far seems to hold up better. It is important to lash the frame rather than screw or bolt it together, as the lashings allow it to flex gracefully without breaking (screws might hold it more stiffly at first, but would eventually snap). The plastic is held on with wooden battens (I found wooden furring strips to be the most economical choice here) and then a piece of clothesline zigzags over the top of the structure, tightening down the plastic. This top clothesline is essential, because it removes the slack in the plastic that results from the sapling frames being slightly different sizes and shapes. Greenhouse plastic is very tear resistant, but it wears out quickly if left to flap around in the wind, so tightening the plastic this way is the key to its longevity.

The original large greenhouse had regular doors, but they were fairly quick to rot and were tricky to make so that they operated smoothly, so when I rebuilt the structure, I started experimenting with a shower curtain–style door that would be easy to make and simple to use. The end walls are framed with two leaning pieces that meet at the peak to form a triangular brace that helps keep the structure stiffer in the wind. The plastic wraps around from either side across to the opposite edge, helping to eliminate drafts. In the summer these curtains are tied open until the late fall when it is cold enough to close them. I used to be much more finicky about opening and closing the greenhouse every day to trap heat and release it, but over the years I started experimenting to determine just how little I could do and still have a successful tomato harvest in the summer and spinach and carrots in the fall. I found that if I waited to plant tomatoes until after the threat of frost, I could just leave the greenhouse open all summer. The tomatoes appreciated the extra heat and relative dryness (although I water them with a tall sprinkler, not ideal but a real time saver), and as long as I give them enough space and water they turn into 6-foot-high (1.8 m) columns of tomato-producing madness.

Mid-July is when I start the carrots I want to harvest in the fall and early winter in a separate area, and at the end of August I sow any spinach, lettuce, cilantro, and Asian greens I want to grow. When the hard frost kills the tomatoes in mid-October, I cut down the tomato plants, unbolt the greenhouse sill from the four stakes driven into the ground, and move the greenhouse over to cover the carrots and greens. In November and December, we eat all the carrots and harvest the greens down to nubs; by then it gets cold enough outside that I move my spoon-carving stump into the greenhouse.

Here is where the greenhouse really starts to earn its keep in a way I never anticipated. The greenhouse gives me a warm place to do the axe work on sunny winter days and at least out of the wind on cloudy days (more on how spoon carving fits into the farm schedule in chapter 7). This makes it possible to continue carving spoons right through the winter, as I can bring the axed-out spoon blanks into the kitchen to

The rebuilt, smaller hoophouse earns its keep all winter as a space to axe out spoons.

carve them. It also gives me the ability to teach spoon-carving lessons during the colder months when I would otherwise not be able to. Our house doesn't have a basement that can be used for this (an otherwise good solution), nor do we have a garage (the other common solution). Our floors are not strong enough that I would feel good about using an axe and stump in the living room, and axing outside when it is bitterly cold and windy is not feasible. I did it for years, but there were days when I didn't make anything because it was too cold.

Someday I will build myself a shop for just this purpose. I know where the shop will go, but I want to wait another three years or so until we are able, financially, to do it right. I have enough appreciation for impermanent solutions to recognize when a permanent solution is appropriate, and this is one such case. I plan to carve spoons for the rest of my life, and I want a structure that will allow me to do so gracefully as I age. It will be a simple space, probably 16 × 20 feet (5 × 6 m), with a very simple layout (a lesson from the You-Cut hut) that could be adapted to a range of purposes. It will be insulated, wired for electricity, and heated by a woodstove that can burn all the scrap wood that I produce.

But for now the greenhouse gets turned into a workshop every winter, with wood chips piling up over the spinach, most of which won't survive the traffic. The lost value of the spinach is made up many times over, however, by the value of the work I produce and the lessons I teach thanks to this impermanent structure. On sunny days, it can be below zero outside and windy enough to shake the sapling frame, and I will be inside, working in a T-shirt, with music playing. This is the beauty of the impermanent solution: It may not be forever, but it can be wonderful in the moment. It acknowledges that both our lives and our needs change in ways we cannot anticipate. Why pretend otherwise? Take off your jacket, crank up the music, and listen to the winter wind roar.

CHAPTER 5

Buying (or Not Buying) the Farm

W e came very close to buying our farm. For a long time we assumed that we would; it seemed inconceivable that we would not eventually own it, and we had no examples of how it might work otherwise. The idea that to succeed in taking something on, whether a farm or any other business, you must own it is baked into American culture. To do less than that is to fail. This, I am here to tell you, is a myth.

Still, I doubt we would have taken over the Christmas trees had we not planned on one day owning them. Christmas trees are a long-term investment. Often the work you do today doesn't pay off for seven to ten years. In such circumstances it's best to just ignore the time frame and do what needs to be done. On properties where the primary crop is woodland for timber, the return on investment is forty years or more from a seedling. We took over the trees in part because we fell in love with the house and the land and knew that to buy them would require taking over the trees. The house was in that early stage of falling apart where nothing is so far gone it isn't worth saving, but everything needs work. The land was a mix of pasture, woods, and Christmas trees. There were two large barns, both in need of some major repairs, and several smaller ones that could be safely ignored for the time. South-facing slope, quiet road, ten minutes from a couple different towns . . . it was a wonderful place.

But one of the obstacles toward purchasing it was that Al Pieropan subdivided his original 25 acres when he built the other two houses, and while the lot divisions made sense in terms of road frontage and optimal house sites, they made absolutely no sense in terms of the Christmas trees. The 25 acres were divided into seven lots, with trees on at least part of every single lot. To make matters worse, the biggest grove of trees, the You-Cut grove, was divided evenly into three lots running side by side up from the road in long skinny strips, with the middle third belonging to the house at the very top of the slope. If we wanted to truly control all of the Christmas trees, we would need to buy not just one house, but two or even three. It would be a lifelong labor of love to pull that off, and one that might never make financial sense. It would also mean we would have to be landlords, something neither of us was keen on after hearing some of the stories Al had to tell.

The groves of Christmas trees sprawl across all seven lots that comprise the original property, which includes three houses, making it an expensive and complicated proposition to buy just the farmland.

Despite this, we really wanted to buy the house we were renting and the 12 acres it came with. This would have given us control of about half of the trees, and it was a nice contiguous block of land, with the pasture, some woods, Christmas trees, and the barns. For a number of years this was our pipe dream. We knew that no one would give us a mortgage until we were better employed, or at least could show more years of income from the farm.

Al was also interested in us buying the farm, and our conversations on the subject grew over the course of three years from guarded agreement based on mutual interest to more candid discussions of how a sale might actually work. We realized early on that it would take several years to get our financial ducks in a row. I had just started an online scientific editing business with the help of my father, and I knew from talking with various banks that they wanted to see three years of taxes before they would consider income from any business when determining mortgage eligibility. We also learned that we didn't have enough streams of credit, as we had only one credit card (which we pay off in full each month) and no car loan or student debt. Mortgage companies base much of their decision to lend or not lend on your credit score, which is a combination of both how much you borrow and how well you pay it back, as calculated in some arcane way by three credit-rating bureaus. We paid money back quite well, but, as it turns out, we were borrowing too little, or rather from not enough sources. Cecilia had cut up her credit card several years before, and we had worked very hard not to go into debt, tightening our belts when money was scarce. But we discovered that to increase our chances of getting approved for a mortgage, we needed to get a second credit card and use it occasionally, to demonstrate with that second line of credit that we could handle borrowing money like adults.

We also discovered that our local banks, recommended to us for a simple, thirty-year fixed-rate mortgage, were tied up in knots about approving a loan for a farm. Or a property that had the ability to be a farm. Or a property that even had the appearance of a farm (no joke, one person told me they would even have problems if there were too many fruit trees). The people I sat down with (a humiliating experience if ever there was one, to lay out your financials for someone and to be found lacking) had no idea what to do with a farm. It didn't

We had grown quite attached to the house and land,
planting fruit trees and berries, repairing and maintaining
the house and outbuildings, and amassing a dream woodyard.

fit into the tidy forms they filled out as we talked. It worried them. It seemed risky, that someone should own land and actually use it for some purpose related to money. Again and again, I was told that there was nothing they could do for me.

Finally, someone told us to check out Farm Credit East, an agricultural loan and farm credit company. Talking with them was a breath of fresh air. Everything that didn't make sense to the bankers made total sense to them. They sent a guy out to talk with us, and he took one look at the place and said, this looks like a perfect application of what we do. Finally, we thought. We could see a path forward.

The Problem with Negotiations

We still had to actually agree on a price with Al, and there we were in for an unpleasant surprise. To avoid negotiating over price, we had proposed (and Al and his daughter, who had power of attorney, agreed) that we would each have the property assessed independently, and that the price would be the middle ground between those two numbers. An assessment is different from an appraisal. An assessment is made by an assessor, and is a dispassionate tallying of a property to determine tax valuations using a fixed formula. An appraisal, on the other hand, is what a real estate agent *thinks* you should start with as an asking price for a property. They're not the same thing, because an appraisal is always optimistically high so that you have room to negotiate down. We deliberately agreed to have the property *assessed* because we wanted to come up with a number without such negotiations. However, while we got the house assessed as we had agreed, Al instead had it *appraised* by the real estate agent who would be representing him in the sale. This appraisal was $50,000 more than our assessment, or $260,000 versus $210,000.

Once we had gotten over our dismay, we decided to proceed anyway, and propose to Al a price that would split the difference between our assessed value and his appraised one. We were shocked and disappointed when he told us that actually he liked the price his agent had given him and that's what he wanted the price to be.

At the time we felt like Al thought he had us over a barrel. We had invested so much in the Christmas trees, and we had done so much to improve the house (painted almost every room, removed several truckloads of trash from the basement, whitewashed the foundation walls, and made plumbing repairs, electrical repairs, roof repairs, stonework, and landscaping) that it was obvious that we *really* wanted to buy. While he was generous in many small ways, it seemed like Al just couldn't stop himself from taking advantage of us when it came to such a large amount of money.

More recently, I have come around to the view that the issue was a universal one, that we tend to overvalue what we have. Al had been given a big number by his real estate agent, and he liked the sound of it. All of his calibrations of what was fair thereafter were based on that number, rather than using it to come to the middle ground as we had agreed.

With my father acting as a level head in the room, we sat down to negotiate further with Al (and his daughter, although she was simply there as a witness and took no position other than to support Al's wishes). After a grueling hour of hurt feelings and mild panic, we came to a verbal agreement for $240,000 (not halfway between but close), at which Cecilia (who had been mostly silent) declared that she felt like she was going to throw up and left the room.

I found her pacing outside, pale-faced. She couldn't do it, she said. It was too much for a house that would suck down every last penny we had and still be hungry for more. It was too much for a situation that would mire us in similar processes of buying the remaining houses and land from Al over the coming years. It was too much to pay.

Cecilia was experiencing the inverse of what Al had, in that having grown accustomed to the lower price given to us by our assessor, any substantial increase over what had come to feel like a normal baseline felt like an affront. Our hope in arranging to have two assessments made was that they would come out to be within $20,000 of one another, which seemed like a reasonable expectation given the formulaic nature of reaching a valuation. The problem for both Cecilia and Al was that the spread between our assessment and his appraisal was so wide that the middle ground felt like too much of a stretch. This is not usually a problem with buying a house, because the asking price sets the expectations, and negotiations flow from there. I still think using the middle ground between two assessments is a good approach for structuring the price when transferring ownership of a farm from one generation to the next, but if the seller gets an appraisal, instead, then I think that what we experienced is almost guaranteed to happen.

In the end, we told Al we needed to sit on it and not make any decisions for a couple of months. This turned out to have fateful consequences, because later that summer a natural gas pipeline was proposed that would run under the power lines just ½ mile (0.8 km) down the road.

In the summer of 2014, Kinder Morgan, a large energy conglomerate, sought approval for a pipeline that would carry natural gas from an existing line in New York out through Massachusetts. The proposed route would largely follow high-tension power lines through what was otherwise extremely rural land. There was immediate widespread resistance by landowners whose land was actually at stake, but also by the general population of the affected towns and region as a whole, as such a pipeline would change the character of the landscape and carry significant environmental risks. These risks were both of the

The proposed pipeline would have passed along the field just behind this barn a quarter mile up the road.

catastrophic kind, should some rupture occur, but also more insidious, as the recompression stations that would be needed along the route carry health risks to the surrounding population.

One such compressor station was, for a time, slated to be built where the power lines crossed our road. Proximity to these stations has been correlated to a host of health issues, including sinus problems, headaches, and skin and throat irritation, while the specific volatile organic compounds emitted by these stations have been associated with a number of cancers, cardiovascular and respiratory illnesses, and even birth defects. Right around when all of this was hitting the news, a compressor station blew up in Pennsylvania, just the latest in a string of explosions to have occurred at such compressor stations.

The possibility of the pipeline changed everything. Any advantage to buying the farm disappeared, as property values would plummet if the pipeline went through. What is more, we were clear that we didn't want to live near a recompression station and jeopardize the long-term health of our family, and we weren't even sure if the road would feel the same, or if the traffic that building the pipeline and the station would entail would destroy the very feeling that we loved about the place. While opposition to the pipeline was high, it seemed to me that whether the pipeline happened or not would be determined by the economics of natural gas, and not by lawsuits or regulation. We told Al that we wanted to put buying the farm on hold until the pipeline was resolved one way or the other.

The House

Over the years, we had routinely perused the house listings for our region, and had even gone so far as to look at four different properties. One was just forest, crisscrossed with stone walls. The other three were houses: one a bizarre home built by an artist, one a delightful shingled cape with a working fireplace and a screened porch, and the third a quirky, small blue house nestled among bedrock outcroppings. All were farther away from the valley and our families than we desired. None seemed like a better choice than the farm. Even so we kept looking, scouring the listings every couple of months for anything new.

The winter after the pipeline was proposed, we were driving to Cecilia's childhood home on Christmas Eve morning when we decided to take a little detour to look at a house we had seen listed. It was a nice size, and had been recently gutted and only partly refinished, which seemed like a good opportunity to complete the interior the way we wanted. When we drove past it, though, I knew immediately that it wasn't the right place. It was right on the road, next to a row of other houses, and while it came with a couple of acres, the house backed up against a steep, north-facing hillside, which blocked much of the sunlight.

"No way," I said as we drove past. "Nope, I am not living there."

"It's not so bad. You always say that. What is it that you want in a house anyway?" Cecilia asked.

"Well," I said, as we continued up the street, looping over to the main road that would take us to her parents, "I always thought we'd live in a house that had better sunlight, if not south facing then at least not north facing like this one. I always thought we'd have a bit more land, and maybe a barn or at least a shed." As I finished talking, we pulled up to a stop sign at the edge of a town common. Across the common, a white farmhouse with a giant sugar maple looming over it had a for-sale sign in the yard.

"What about there?" Cecilia asked. "Would you be happy there?"

"Yes," I said, without hesitation. "Why yes I would."

We had not seen the house on any of the listings, and it turns out that this was because the asking price exceeded the parameters we had been searching for, and while we had searched for more expensive homes that were listed as two-family properties (we had a dream of taking our current housemate, Melissa, with us), this house was listed as single-family and so it had slipped through the cracks in our searches.

It had a front and side porch, white clapboards, and a slate roof on the main part. A small red barn sat off to one side. It was situated on a narrow triangular common, with two more houses on the other two sides. This common made a huge difference in the feel of the neighborhood, giving the houses some common ground, and space from the others. It was ten minutes from the farm, and just ¼ (0.4 km) mile from the main road going down to the valley. The location could not have been more ideal.

The new house from across the town common, the same vantage point that we first saw it, although it would have been winter then.

After Christmas I drove by in the mid-morning, and was shocked to see sunlight playing over the entire face of the house at ten a.m. on one of the darkest days of the year. At the farm, our half of the house faced north, and we had just one upstairs room that got direct sunlight during the winter. The thought of a house where every room was awash in light during the winter made me giddy with delight.

We immediately realized that for the purchase to work with our finances, we would need to be able to split off a section and make an apartment. Renting the apartment would bring the mortgage down to a level that we could afford. We asked our current housemate Melissa if she would move with us if we managed to buy the place, and she said yes. So when we had our first viewing, we were hoping to determine if there was an easy way to create an apartment.

Our initial impressions of the inside were mixed. It was hard to see past the wall-to-wall antiques the owner had amassed in every available space. She ran the house as a bed-and-breakfast, which meant that it had a strangely unlived-in feel, overly formal. The ceilings were enormous, 10 feet (3 m) high, with windows to match, and the front hall had a curving staircase and banister. Our friend Michael, a builder, had come along to advise us. When I remarked that the ceilings might be a bit grand for us, he shrugged and said "give it two weeks." He was right, in the end; the high ceilings are now one of our favorite things, and spending time in a house with normal ceilings feels strangely cramped.

The house was built in 1870, which was the sweet spot for passive solar design in vernacular American architecture. In the 1830s, rolled sheets of glass became cheaply available, making it possible to easily create large windows. Electricity didn't reach cities until the turn of the century and likely didn't reach this house until the 1920s or 1930s. So for a century or slightly less, builders had the means and the incentive to build houses with windows far larger than were used before or after. Almost every single window in our house is 3 × 5 feet (1 × 1.5 m), which is enormous. The interior of the house is awash in light, a feeling amplified by the high ceilings.

The house was not without its problems. During our first visit, there was an inch and a half of water in the basement because somehow the sump pump got unplugged. The floor in the kitchen was squishy and some minor repairs were needed on the exterior. But on the whole, the house was sound, strongly built, and we could move in and pick away at things without having our lives consumed by renovations.

In the end we managed to buy the house. We figured out a way to install a kitchen for us in what had been a formal front parlor, creating an apartment with the original kitchen at the back of the house for Melissa. We managed to secure a mortgage (the extra year between

Homebuying

I am not an expert at buying houses. I have only done it the one time, but here are the useful tips gleaned from our experience. This is in no way meant to be comprehensive.

Hire professionals. While you might be tempted to negotiate a sale alone, having an agent act as a middleman can be helpful when things get heated and feelings start to get hurt. There is a good reason why it is most common for both the buyer and the seller to have agents. Having a middleman saves you money in the end and can help prevent deals from falling through. We also found it helpful to hire a mortgage broker to find us a mortgage. Because most of our income comes from being self-employed, traditional banks were leery of lending to us. Our mortgage broker was on our team, advocating for us, and they really earned their fee. Finally, our real estate lawyer proved crucial at the last minute when the mortgage company demanded additional documents as we were sitting down to sign the closing documents, and she basically told them to get lost. Having people whose job it was to go to bat for us when things got rough is the *only* reason our deal didn't fall through multiple times.

Calculate the actual cost of living in the house and compare that with the cost of where you live now. This will help you make a more informed decision about whether or not you can afford the house. In our case, with the apartment income factored in, our monthly expenses living in the house were projected to be slightly less than it was

when we started seriously contemplating buying the farm and when we saw the house we bought turned out to be really helpful, because each year we made more money from the businesses), and we negotiated a price with the seller.

costing us to rent our half of the farmhouse. Once that became clear, the decision to buy was an easy one.

Put as much money into a down payment as you can. Obviously you should compare your mortgage interest rate to the rate you would pay on loans if you anticipate needing to do major renovations, but often these loans can be part of the mortgage. And while it might be tempting to compare the return you could get by investing more money rather than paying down your mortgage, there is no guarantee of any average return, whereas paying down your mortgage is a sure deal. Putting more money down also eases your monthly payments. You should still keep a six-month supply of cash in the bank rather than pour it all into a down payment, but beyond that, putting down as much as you can afford will mitigate as much uncertainty as possible.

Make a thirteenth payment every year. This will help you pay off your principle early. Make sure you are allowed to pay down the principle without penalty, something that is pretty standard for fixed-rate mortgages, and then push yourself to do it. Over the lifetime of a thirty-year fixed-rate mortgage, you pay almost as much in interest as you do in paying back the principle (or more, depending on the interest rate you get). Paying down the principle even a little bit will save you thousands of dollars in interest that you would otherwise have to pay.

Maintain your house once you buy it. Remember that for most of us, our house is our biggest asset. As such, it is worth pouring money into maintaining and growing its value. Even if it is the home you want to stay in forever, and you want your family to own forever, be prepared to pay thousands of dollars a year to maintain and repair it. Even if you do most things yourself, a house costs money.

My brother was instrumental in helping us weigh the costs and benefits of buying this house versus buying the farm. I was in a panic that we were making a huge mistake, that we couldn't afford to buy the house (it was more expensive than the farm, though it didn't need the extensive repairs that the farmhouse did). With my brother's help, we were able to accurately determine what it would cost to live in one scenario versus the other, and how easily we could regain our investment should we need to sell. One of the big selling points of the house was that it was in a wonderful location, and was just a house, with no complications like a farm sprawled across multiple lots. Should we need to sell it, we could do so easily. In addition, the extensive renovations the house had undergone ten years before were now part of its value. It is rare for the value of a house to increase by an amount matching the cost of a renovation. With the house on the farm, even if we had spent $100,000 fixing it up, it would not gain that much in its value. With this house in town, the renovations were done and the value adjusted to take this into account.

By sinking every penny we had or could scrape together into the down payment, we were able to buy the house. The day after closing I started painting; in two and a half days, I painted the entire inside. I also cobbled together an unfitted kitchen in the front of the house for us, using two premade islands, a 2 × 4 frame covered with butcher block for the sink counter, and some open shelving, and we moved in two weeks after closing.

Farming from a Distance

I wasn't sure how it would work to commute to rather than live at the farm, but it turns out it was easy. The distance has actually been helpful at times, giving me the mental space to think about the farm more clearly as a business. Farm decisions became business decisions, separated from home decisions. There is an infinite amount of work to be done on any farm; by not being there 24-7, I stopped being reminded of this work every day, which clarified what work mattered and what didn't.

Not living on the farm also helped me realize how much it had contributed to a kind of tunnel vision, limiting my view of all of life's possibilities. If anything, moving off the farm helped me think bigger about how I could leverage the skills and experience from the farm into other opportunities. I would not be writing this book, for example, without the broader view of my life that moving off the farm gave me.

The picket fence out in front of the new house is the perfect spot to sell some trees and wreaths to locals driving by.

Even before we moved, Al sold the farmhouse and land that we had considered buying, and was waiting to close as soon as we were out. Once we were out of the old farmhouse, we spent a day deep cleaning every surface and corner in the space. It felt like good karma, and it felt like saying a proper goodbye to the house that had been so important to us, where our children had been born, and where we had become adults.

We didn't know if the new owners would want us to continue farming the trees on their land (about half of the 10 acres), but they turned out to be delightful, appreciative of the house and the land and happy to have us continue tending the trees. Two years later, Al sold the house at the top of the You-Cut grove that controlled the middle third of that section of trees, and we found the couple who bought that house were also on board with us continuing to farm. Our fear of losing control of the trees if we didn't buy the land has so far proven unfounded. Only 3 acres of trees remain in limbo, and it seems pretty clear that there is enough momentum for us to be able to ultimately secure leases on these remaining parcels if the property leaves the Pieropan family.

In retrospect, the very thing that made it difficult to buy the Christmas tree farm—the way the trees were scattered across lots and all mixed up with houses and the landscape in general—ultimately made it possible for us to successfully lease it from multiple parties. When the parcels were for sale, anyone considering buying one knew that some of the land had a lien on it because of the Christmas tree farm; indeed, multiple prospective buyers lost interest in the house at the top of the You-Cut grove for precisely this reason. No one controls enough of the trees to make going into business side by side with us worthwhile, and so the easiest thing for everyone is to just let us keep doing what we're doing.

It also undoubtedly helped that trees are a perennial crop, an almost ineradicable part of the landscape. If you cut them down you would not get pasture; you would get an unmanageable sea of stumps, fast sprouting up to hardwood forest. So short of trying to farm the trees themselves, the new owners have little incentive to use the land for some other purpose. In this regard, pasture or crop farming carries more risk of a lease falling through than an orchard. The more changeable the landscape, the more tempting it will be for owners to reclaim it and use it themselves.

Leasing the Farm

Our leases with the new owners are much more straightforward than what we have with Al. Remember that with Al, we paid a percentage that decreased each year and bottomed out with us paying the equivalent of property taxes. The new leases stipulate that all sales of greens or wreaths are ours, and that we pay 5 percent of the sale of trees to the owners. The beauty of the flat rate is that it fairly compensates the owners as we increase production and their cut gets bigger. It is a fairly small percentage because we were, quite frankly, willing to walk away if we had to negotiate higher. Having spent years paying hefty percentages to Al, we were unwilling to pay 10 or 15 percent for trees that we had worked so hard on. Five percent is enough to acknowledge that the land is not ours, and comes out to several hundred dollars. It is not meant to make serious money for the owners. We also try to be respectful tenants, communicating actively about our plans and what we need in terms of truck access. We do our best to improve the land each year, clearing multiflora rose hedges, keeping pasture mowed, and generally leaving things better than we found them.

Our lease is for ten years with the stipulation that we can renew at the end of this time. We chose not to pursue a longer lease because backing away from trying to buy the farm made us feel like our lives were opening up to new possibilities. And while I certainly intend to farm the trees until I'm old, I am also aware that life is long and full of unforeseen changes, and it doesn't seem out of the question that I might want to pass things on before then. Keeping the lease short allows me to think clearly about the opportunity costs of farming versus doing other things with my life, and helps me feel like if it makes sense to move on, we will.

Buying a farm is wonderful. It gives you greater control over what you can do with a piece of land, and generally this leads to a deeper relationship with the landscape. But buying a farm can also be a trap, a burden, or just not the right move. It is okay to not buy the farm. Sometimes you need to walk away. Other times you might find that you can have your cake and eat it, too. You can decide to not buy the farm but continue leasing it. For us, this has been an unqualified success. We love our new home. We love that our kids have friends right next door. We love that we live ten minutes closer to our families and to the towns we frequent. We love that we are still just ten minutes from the center of Ashfield, the town where our farm is located, the same distance it was from the farm. We love that we were able to continue living with Melissa for several years. We love the tall, sunny rooms and the neighborhood and the fact that our acre and a half is enough to feel varied and spacious but not daunting. We love that our new barn is tiny and not in need of any repairs.

Leasing the farm has had its moments of worry, but now that things have settled, everything has worked out about as well as it possibly could have. We love the new owners. We continue to feel a deep commitment to the land and the trees and to carrying Al's work forward to the next generation. We find it easier to make good business decisions when the land is distinguishable from our home, and it is easier to follow other life pursuits with fewer of the responsibilities that buying the farm would have entailed, like fixing the house, repairing the barns, and keeping the fields mowed. I get to arrive at the farm, take care of the trees, and then go do other things.

America is the land of ownership (or at least we like to think so, even as fewer and fewer of us technically own our land, our houses, or even our cars). We worship the idea of buying things, in part because once a purchase has been made, the use of that purchase feels like getting something for free from then on. It's painless. You buy the pickup truck and it feels like you're getting your money's worth every time that you pick up some lumber, even though you can rent a pickup by the hour or by the day from any Home Depot for much less than owning one. Buying land can feel like an investment in the future, and sometimes it is indeed the right move; sometimes it is the only way you can gain enough security to pursue the farm you want, as there are many players willing to lease the same land. Under such circumstances it makes sense to buy, as long as you can afford it and have a real plan for the farm.

But many times it makes more sense to lease. To rent. To go with the impermanent solution, because you never know how things will end up or what you will want to do in ten years. Sometimes it works out better that way, even though we don't have a national narrative around leasing. When you own your land, you can tell everyone to buzz off. You can be ornery and rude. When you lease, you need to maintain good relationships with your landlord. You need to be good at communicating, and diligent about meeting your commitments. While it might be tempting to think of the joy of being a curmudgeon if you owned land, the skills it takes to succeed as a tenant are exactly the same skills it takes to succeed as a farmer: sociability, communication, and reliability. And who knows? You might actually prefer leasing.

CHAPTER 6

Growth and
Its Discontents

F armers think about growth a lot. We plant seeds in the spring and plot their trajectory to harvest. We watch baby animals turn from bits of fluff and twitching tails into majestic live-stock. We monitor the yearly surge in our trees, watching the tender shoots harden and mature after the reckless push of spring. We understand that some growth is cyclical while other growth builds on itself year after year. We spend time calculating the carrying capacity of a meadow, or a barn, or a row of seedlings. What is harder is matching the farm's growth to that of your dreams. Sometimes you dream big but have only a certain amount of space with which to work. Other times you dream too small and struggle to make ends meet.

We have expanded and contracted several times over the years, in several ways. In the beginning, we were entirely focused on growth, as there were always more overgrown sections of grove to rehabilitate and I had nothing else to divide my time. Year after year we sold more and more trees from the You-Cut grove, and we started taking over the remaining sections of trees from Al after a couple years. Word of mouth and our website and flyers led to new customers, a high percentage of whom became repeat customers.

We've seen similar growth in the sale of our wreaths and greens. For the first few years, our wreaths sold almost entirely at the farm, to only one large account. But when some of the larger CSAs in the valley started building farm stores, and wanted wreaths to sell over the entire holiday season, the number of wreaths we made more than doubled over the course of two years.

After my initial excitement at finally having some money coming in, it became clear that my body couldn't keep up with the pace. An old injury flared up and I started getting numb hands from squeezing clippers all day, harvesting greens and breaking them down for wreaths. I would wake up at night with my hands completely asleep, and would have to shake them and beat them on the bed to get the blood recirculating. Clearly, this was an untenable situation.

At first we thought an assistant might help, so we hired a friend to come and cut greens a couple days a week while I tied wreaths.

The truck rack is specially designed to handle two tiers of wreaths, allowing me to deliver up to seventy-five wreaths in one go.

However, it didn't really solve the issue because I still had to use my hands to tie wreaths all day and to harvest greens on the days my friend wasn't there. Since cutting the branches off the trees required the largest cuts, whereas breaking the branches down into the appropriate sizes required relatively small, easy cuts, having someone just helping to break down greens that had been harvested was missing the point. I tried to save my hands using a billhook, a sort of short, heavy machete, but even though the billhook allowed me to keep my hand more relaxed, it was slower and less versatile than clippers, and in the end I returned to using clippers. Similarly, I tried using much larger clippers rated to a wider-diameter branch, but found that not only did the larger size require more effort to squeeze shut, they encouraged me to cut larger branches than I otherwise would, further exacerbating

During such a busy season, you do what needs to be done, which sometimes means working in wet, cold conditions.

the issue. Cecilia keeps suggesting I try getting battery-powered pruners that would do much of the work for me, but I suspect I would compensate for that, too, just as I did with the larger clippers.

In the end, two things unrelated to technology have helped my hands. The first is that increased spoon carving has left my hands stronger and more resilient. Instead of going into the Christmas season with relatively weak hands and building the strength over the course of a couple weeks, I essentially train year-round to be able to do this marathon injury-free. This should be common sense and come as a surprise to no one, and yet it was a surprise to me.

The second thing that continues to help—and this is the kicker—is to do less. No matter what else I've tried, whether hiring extra help, equipment hacks, or any change in operations that gives me a break, I've compensated by doing more. Over the last three years we have hired significant amounts of help, both full- and half-time employees; if anything, the responsibility of having to pay someone made me take orders that I might otherwise turn down. In the end, I found that doing less has had to come from within, and from saying no to opportunities rather than figuring out a workaround.

Employees

Over the years we have hired a number of people to work for us, usually for just one season. Sometimes we get along famously and chatter away for hours, and sometimes we just plug in our earbuds and listen to our own music or podcasts. Training different people has made me aware of how much nuance goes into this work, and how fast I have gotten at it. Some people, we've found, have the knack for working swiftly and smartly with their bodies, on practically a microscopic level. Others do not, and it is almost impossible to tell which is which until they are in it.

With the right person, the days fly by and the work seems less. With others, you mostly ignore each other, which can make the days seem longer because you also don't enjoy the complete freedom that comes from working alone. I quite like the balance of having help for half the days of the week and then working alone for the other half, but it does mean that there is less flexibility to adjust the schedule

Crew members Maria Darrow and Kyle Farr overlooking the lower field and farmhouse. Just behind them is the stack of cut trees we just hauled to the edge of the driveway and staged for loading.

around the weather. When I have full-time help (or no help), we work outside in nice weather when jobs like cutting and hauling trees to their staging points are best accomplished, and if it's cold and rainy or snowing we work inside tying wreaths. But when I have help only half the time, I sometimes end up needing to do two-person tasks such as cutting and hauling in bad weather, simply because of the other person's availability.

As a boss, I have tried to follow the example set by my favorite captain from when I worked on sailing ships. Captain Ryan was generous with praise and gratitude, read the pulse of his crew, cheered us up when our energy was flagging, and while he was never afraid to be goofy was all business in a crisis. I am grateful for the chance to try to emulate him, and I am grateful for the hard work our crew has put in each year, but I have come to the realization that at least for the next year or two, I need to step back from employing others and just work solo.

Being a Good Boss

Being a boss is one of the most rewarding things in life if you approach it with the right attitude, and one of the most frustrating if you don't. Here are the things I keep in mind when I'm in charge of other people.

Communicate constantly. Give both positive and negative feedback. Communicate your gratitude, and immediately address things you would like to be different, before you have time to dwell on them and get resentful or grumpy or dread having the conversation.

Set the tone. Remember that you're in charge of the overall mood! Bring special snacks. Choose good music to listen to that will bring people together. Don't be afraid to be goofy.

Set the pace. Don't expect other people to work harder than (or even as hard as) you. After all, it's a job for them, but it's *your* life. Instead, set an example by being an absolute ninja at what you do. Never ask anyone to do something you wouldn't ask of yourself. Share the tough jobs and the fun jobs.

Take care of people. Send them home if they're sick. Let them leave early if they need to. But demand clear communication. If they need to arrive late for some reason, that's fine, but they need to let you know in advance or move heaven and earth to arrive on time. Their role is to take the job seriously. Your role is to be flexible when possible.

Pay people on time. I will admit that I have not always been good about this. Not because I didn't have the money, but because sometimes I didn't submit their hours to the bookkeeper on the day I should have, and that is unacceptable. Employees are people who exchange their time and effort for your money. Treat this exchange with dignity by being scrupulous about paying them on time.

Pay

Part of why our farm needs to step away from hiring help goes back to the idea of doing less, and part of it has to do with profit margin. We start pay at $20 an hour, a high wage in farming. The one time we had someone come back a second year we paid her $22 an hour and offered her a cut of the business to stick around after that, although she ended up starting her own venture elsewhere. One of the reasons we pay so well is because we are offering work for only a month and a half, so we need to entice people to disrupt their other opportunities to work with us. But we also pay high wages because I demand a lot from people. Twenty dollars an hour is real money, and I require real effort in return. Our work is fast-paced, detail-oriented, and sometimes unpleasant.

The third and most important reason we pay as much as we do is because $20 an hour, where we live, is a living wage. A real, afford-your-rent kind of wage. I have a big problem with the low wages agriculture as an industry pays its workers. Even groovy small farms pay a pittance, justifying it by calling it an internship or apprenticeship, by saying they can't afford to pay more, or by pointing to their own tight belts as proof that they are being fair. There is nothing fair about employing someone for an hourly rate that damages their own ability to succeed in life. If you can't afford to pay someone well, figure out how to farm so that you don't need help. Earn less yourself. You own the business, and are building equity and opportunities in a way that an employee never can. You are in charge!

I worked for years on both sailing ships and farms that paid very little, and as a result my own sense of self-worth was skewed for a long time, in ways that impeded my ability to provide for my family. I just didn't value my time at $20–$50 an hour the way I do now. Until someone else tells you that you are worth that much, you don't believe it. Partly I felt that because I was young. Is a twenty-year-old's labor worth that of a thirty-year-old, all else (responsibility, temperament, work ethic) being equal? Not in my view. Experience and maturity are worth a lot. But when it comes to hiring for our farm, I feel strongly that by paying $20 an hour for our help, we raise the bar on our employees' self-worth, pay fairly for a fast-paced, brief window, and push back against the low standards of our industry.

There are other costs to hiring someone in addition to their wages. In the state of Massachusetts, when we employ someone for more than $600 (the limit for "casual labor"), we need to pay half of their Social Security payments (the other half gets withheld from their wages), hire a bookkeeper to manage the system (although honestly this cost is minimal and our bookkeeper, Kara, does a phenomenal job), and purchase Workers' Compensation and Liability Insurance.

So at the time of this writing, I'm planning to go solo in the coming year to see if running a smaller, more streamlined operation will be more profitable. On paper I should save between $6,000 and $8,000 in costs, while losing much less than that in revenue. My hope is to net an additional $4,000 to $6,000. The big question is whether my body can hold up, although this last year leads me to believe that it can. It was the first year that I didn't wake up with numb hands, and I have been carving even more this year than the previous one, building the hand strength that carving demands.

Limits

The other reason I'm going solo is to make the farm more profitable on a smaller production base. Because the amount of land in trees is fixed, there is a real limit to what the farm can produce year after year. While plenty of land still remains to be brought into stronger production, this takes time, and we are near capacity for areas that can be improved quickly.

At this point, there are three main reasons why some of our trees become overgrown. The first and most noticeable is color. Certain areas of the farm suffer from poor drainage, resulting in yellowing of the balsam needles starting around Thanksgiving. Adding nitrogen had no discernable effect on this yellowing, although it did double the amount of growth produced. As mentioned earlier in the book, I intend to supplement a test plot with iron to see if that helps; if it doesn't then it will be time to think about digging ditches. About 4 of the 10 acres are affected by yellowing, although trees and greens can still be harvested for wholesale in November before the yellowing occurs, since the color of the needles remains fixed after harvest. Of those 4 acres, 2 have become overgrown to the point where some sort of intervention is necessary.

After a long day of working with balsam, your hands are black with pitch. The best way to get it off is a hot shower followed by washing a load of dishes.

The second reason is inaccessibility. A number of areas have become overgrown because they are difficult to reach, either because the ground is steep or because they are simply far from any point of access. Taking trees down once they become large, even if the branches can be harvested for greens, requires a chain saw. The smallest chain saw I can find would likely be a better choice than using the Stihl Farmboss chain saw I currently use; another instance where spending $400 on the right tool would save me thousands of dollars in time and effort. But cutting down larger trees reveals an additional hurdle: Big stumps have often aged beyond the point where they can resprout, and so the area would need to be replanted.

For the last eight years, rehabilitating the balsam groves has meant dealing with the low-hanging fruit. Every step of the way, I've asked myself, "What area has the greatest production potential for the least amount of effort?" Then I'd go put in the time there. But at this point, all the low-hanging fruit is gone, and the next several years will involve tackling areas where there is much less to salvage and much more to be done. My hope is to be able to harvest enough greens from this salvage and from the stump rotations of the more productive areas to meet demand, but this potential constriction in balsam supply is another reason I want to make the operation leaner.

Finally, the biggest constraint to growth on the farm is the limited window during which most of the work takes place. For instance, rehabilitation necessarily involves cutting down overgrown trees and clearing paths; it's best done during the harvest season simultaneously with harvesting greens, but this also means there's a limited window for this kind of work each year. Similarly, because balsam harvested before the first couple of hard frosts won't hold on to its needles, there is a limited amount of time during which to harvest greens, tie wreaths, and cut trees. I tend to shift to working full-time in the trees in the last week of October, and spend a week, more or less, cleaning and preparing the You-Cut hut, setting up the tarp barn, spreading wood chips, cutting multiflora rose, and clearing trails. As soon as the third hard frost hits, I drop what I'm doing and start cutting greens. This first date has ranged from October 25 to November 2 in the years I've been doing this, and in a season where all of my wholesale accounts want their orders before Thanksgiving, that extra week can mean the ability to handle several thousand more dollars of work.

There is only so much time—in the day, the growing season, the year—and so the name of the game for working on the land is to prepare as much as possible beforehand. I wire clusters of pinecones for decorating wreaths; I tie as many bows as I can store under the bench (about 25 percent of what I need for the season); I order all my supplies and make sure my tools are ready so that when the time comes to finally cut greens, that's all I need to do.

In terms of squeezing more out of each day, the most recent big example was replacing the sheet metal roof of the hut with polycarbonate paneling, effectively turning the entire roof into one big skylight. This move keeps the inside as bright as the outside, allowing us one extra hour of work each day we are tying wreaths, as we need to stop when we can no longer see the true color of the greens, and the sun sets early in November and December. If we tie wreaths for half of the days between the beginning of November and mid-December, that means twenty-two extra hours of work, or almost three days. Since for each seven-hour workday this time of year I earn at least $350 depending on what I am doing, that extra twenty-two hours means an extra $1,000. The polycarbonate roofing cost $160 and took five hours of my time and $80 for four hours of someone else's time to replace. You do the math.

Scale

One of the trickiest things to balance on any farm is a scale that works for you, for the land, and economically. For most farms, this balance shifts from year to year as the business grows, and it is a better fit at some times than at others. In 2007, Tim Wilcox and Caroline Pam started their farm, the Kitchen Garden, with 1 acre. Since then they've grown in leaps and bounds each year; now, a dozen years in, they farm 50 acres, cobbling together a number of fields in several towns. When I worked for them for a season many years ago (when they were a fifth the size they are now), Tim told me about how growth can affect the economics of small farms. When scaling up to acreages that require switching from hand labor to tractors, or from small tractors to bigger tractors, for example, the revenue from the increased size may not be as much as from an even slightly larger increase in acreage where you would use the same new equipment. Certain large scales also require more bodies, but then that same increased number could handle a further increase of just a couple of acres more. Seven acres could therefore be more efficient than 4 acres, if it can be farmed with the same equipment needed for 4 acres, and only one or two extra people, since in theory you can produce almost twice as much produce.

Price and Growth

My land is a fixed variable, but two major things I can change are the amount of labor I bring to bear on the land in relation to demand for our products, and the amount of growth the trees will sustain (or excess growth that needs to be removed). These last two are moving targets: Demand for our greens and wreaths and trees has slowly risen over the years, while the amount of excess growth that can be easily harvested has declined. For a while it made economic sense to bring more labor into play to handle the demand. At the time of this writing, however, I've decided to increase our wholesale balsam price, which has remained the same for eight years, by 20 percent. This should drive down demand a bit and also make it more worth my time to pursue greens from less efficient areas of the grove, where there will be chain-sawing involved.

In general, we have used price as a means to depress demand, although it is only somewhat effective because there is an upper limit that feels weird to cross, given that we want to keep our trees afford-able. When we took over the farm, we had a non-compete agreement with Al whereby we would keep our tree and greens prices the same as his. Al had been selling trees for a flat $20, but the year before we took over half the farm he raised it to $25, and that is where the price sat for seven years. This last year, with Al no longer managing any of the trees and some concern that we needed to depress demand, we raised the price to $30. No one batted an eye.

Our wreaths started out even lower, and have been more responsive over the years to customer demand. Our strategy is to raise prices piece-meal, rather than across the board, so first wreath prices responded to demand, then tree prices, and now wholesale greens. Wholesale pricing is the most brittle, because businesses lay out more money at once and also analyze their margin. A You-Cut customer just makes a fairly small adjustment in their mind, and they're also usually already at the grove when they realize the price has increased, although we do announce it on our website. So until now we have held off increasing the price for our greens, anticipating that we might lose a few wholesale customers. Now is the appropriate time, however, because the grove needs a couple years of harvesting less of the easy stuff and more of the hard-to-reach stuff, which means there will just be less of it.

I like to let naturally occurring patterns inform the timing of price increases, so they don't feel arbitrary to customers. Telling wholesale partners that the greens price has increased because I've run out of easily harvested greens and need to start harvesting overgrown stuff has a logic to it that increased cost of living or some such justification never could. Most will probably buy into the vision of rehabilitating the grove, and stay. Similarly, with the tree price, waiting to raise prices until the grove was clearly getting harvested heavily helped people value the trees more, seeing the demand with their own eyes.

An Alternative Model of Growth

Cecilia and I met Tim Wilcox and Caroline Pam of the Kitchen Garden when we had side-by-side booths at a farmers market, and we became friends over the course of a season of hanging out and waiting for customers. Two years after leaving our vegetable farm and moving to the tree farm, I worked for Tim and Caroline for a season leading up to our first year taking over the trees. They have grown their farm almost every single year, and as such their example provides a good counterpoint to our own example of restrained growth.

The Kitchen Garden started out on 1 acre, and when we first met our farms were roughly the same size, 3–4 acres. When I worked for them a few years later, they were farming 7 acres and had a crew of six. The following year they entered a short-lived partnership with another farmer that gave them access to equipment and land that allowed them to more than double in size, and now, nearly a decade later, they are farming 50 acres. Tim and Caroline divide the farm into production and harvesting/sales, with Tim managing a team of eight on the production side and Caroline managing a team of sixteen on the harvest side. They have thoughtfully and aggressively used a combination of bank loans and state and federal grants to achieve this growth, and have shown a willingness to invest in technology and systems improvements every step of the way.

It would be easy to look at the Kitchen Garden and see a farm very different from ours in terms of trajectory and scale. But I would argue that our two farms are actually making very similar strategic decisions. In the last three years, for example, they leveraged their interest in hot peppers to start an annual Chilifest and then leveraged that reputation and momentum to start making sriracha and other hot sauces that are sold across the country and have appeared in cooking magazines and in the kitchens of celebrity chefs. While still a relatively new part of their business, the value-added products account for about a quarter of farm revenue and they are investing in building their own certified kitchen space to give them more flexibility than the community kitchen they use now.

Tim and Caroline also offer above-average compensation and have worked hard to create a cohesive farm culture. The crew takes turns preparing a sit-down midday meal to share with everyone (I still dream of the amazing food I ate when I worked for them). It is a testament to their management and leadership that many of their crew have chosen to return for the last five years, and that as the farm has grown they have been able to give people expanded roles with more responsibility.

The Kitchen Garden is extremely diversified in its crop production, specializing in many heirloom varieties that chefs want, and their customer base is built around these relationships. Caroline is a food writer and went to culinary school while Tim is an equally passionate cook, and so the farm is an expression of their love of food.

In terms of nimble infrastructure, before they built large packing and storage space and greenhouses, they retrofitted and utilized some of the existing structures on their farm. When they finally got around to building their current arrangements, they knew what they wanted and needed more clearly than if they had built something at the start. They are also practicing the principle of land rehabilitation with their current construction of a new certified kitchen space; rather than plop it down somewhere new, they are demolishing an old falling-down barn and reclaiming that space.

With my wooden spoon prices (more on this in chapter 7), I've decided that October is the perfect time to raise my prices across the board, as it's when I close my waiting list for the year to pivot to the farm and start taking spoon orders for the following January. That timing allows me to announce it multiple times leading up to the cutoff date, perhaps driving a bit more business from people wanting to get in under the wire, and then by the time I start filling people's orders in January the new prices have been in place for several months and feel normal. Spoon pricing is tied to a long-term strategy that I will lay out in the next chapter, but the plan is to increase prices each

Of course, not everything they have tried has worked out. Over the years, the Kitchen Garden has participated in a number of farmers markets, operated a boxed delivery CSA, and delved into catering. But by throwing a lot of mud at the wall, they were better able to run with what stuck, including wholesale relationships with stores, distributors, and restaurants, and the prepared hot sauces. These efforts provided the best return on effort and potential to grow, but this only became clear because they tried all the other stuff.

Lastly, Tim and Caroline have always been extremely savvy with their use of their website, newsletter, social media, packaging, and brand in general. They use all of these tools to share their ongoing story with customers and are always looking ahead to what changes are coming over the horizon, both externally that they need to react to and internally that they can create for themselves. Tim mentioned in a recent podcast interview that after passing through a period several years back where they were both considering leaving the farm through burnout, they got to a place where it feels very settled and clear that they will continue. This change seems more internal than external, driven by a shift in mindset rather than circumstances. They will continue to grow the farm on a number of different fronts, and while their growth seems steep now, I suspect it will seem like just the beginning when looking back in another ten years.

year, doubling the original price in about five years. By announcing the increase in advance and having a lag time after the increase before I produce any spoons, and by having it fall at a time of year when money seems looser in general (Christmas), this price change has so far been frictionless and largely unnoticed.

Pricing is always a balance between the money you hope to earn and the level of demand you hope to retain or create. Your price is informed by the nature of the business (how much of your business comes from repeat customers?), your long-term strategy (are you building a reputation or skill and using low prices to create demand that you can leverage later?), and your gut (how are you positioned within your community?). I have found success in starting with low prices and increasing them as demand surpasses my ability to keep up. If the time you have to pursue a farming venture or business is seriously limited by another job, however, or if you are extremely skilled at what you do, or if you already have the demand in the form of a strong following in your local community or on social media (depending on which is more applicable), then you might find it a better choice to set high prices. Just don't put the cart before the horse. The trick is to match the price to *your ability to meet demand at whatever scale that is for you.* If you are looking to sell only a few of something at a high price point and the demand meets the limited time you have to put into it, that is probably a good fit. If you have nothing but time and no one knows who you are yet, then use a low price to build a base of customers.

Customer Service

Making sure your customers remain loyal is the key to building price leverage; while this comes down to the quality of what you are selling, equally important is the quality of the service you provide. You need to be growing amazing vegetables, unsurpassed fruit, top-quality meat or dairy, or making really excellent wreaths (imagine that!), but you also need to be diligent about returning every phone call, responding to every email, delivering on time, and accommodating the timing needs of your customers. The ease and responsiveness of working with you is just as important as how good your product is, if not more so.

I will admit that I was not always as good at customer service. It's a commonly held idea about farmers that they are hard to get a hold of, out in the field, slightly taciturn, and a bit awkward from being alone all day. None of these stereotypes are helpful when it comes to building a customer base. You need strategies for keeping on top of customer service.

For me, this started with just a big sheet of paper taped to the wall. As wholesale orders for the trees came in, I would write them on the sheet along with billing information, quantities, delivery dates, and anything else that seemed pertinent. With my spoon carving, I handled orders in the same way when rising demand began to outstrip

Christmas Tree Farm

spooncarving lessons

The lead-up to Thanksgiving involves an awful lot of standing at the bench, tying wreaths as fast as I possibly can. *Photo by Meghan Hoagland.*

my ability to get by with scraps of paper. Pretty soon, however, this system became cumbersome, and I found it easier to enter orders into a weekly planner, which allowed me to estimate wait times on a rolling basis. Just recently, I have switched to a larger planner to give me more room to write down details as I push more work into each day. I like using analog planners because they provide a secure backup to the information that is on my phone. To this end, any information customers choose to give me gets written down with their order, so that I am not reliant on my phone working to know their address when the time comes to box up an order.

Opportunity Costs

One of the trickiest things about growth is analyzing the opportunity costs of one decision over another. Opportunity costs are all the things you *aren't* doing in order to do the thing you *are* doing. For instance, if I say yes to a large order of roping (which has the lowest profit margin of anything I make for the farm), that represents time that I can't spend filling another order at a higher profit margin. But what if the roping order is so large that I'm not sure if I will get enough orders to make up for not taking it? What if the roping order is so large that I hire another person specifically so that I can take it? What are the marginal returns on that? What if I take on a really huge wreath order this year, doubling the number that I do, and jump the business to a whole different size (this offer has been on the table multiple times, by the way), and then next year, I think the same order will materialize but it doesn't?

I also think about this when balancing one of my business's opportunities against another. I could do a lot of holiday fairs selling spoons but I don't, because I will obviously make more money pouring that time into the grove. But what if there was a teaching gig in Australia that would conflict with the Christmas trees? What if there was a book deadline? What if I just had a huge month with the scientific manuscript editing and needed to back off the trees to do that? Opportunities are easy to analyze until your time is full. Then they start competing with one another, and the more moving parts there are, the more complicated the balancing act.

The biggest factor I have found over the years that contributes to growth is consistency. That's really it. Just being present, doing your best, year after year after year. If you use social media, do it consistently. Post every day, or every week, and be disciplined about it. Figure out how you can best bring value to your customers. Maybe that takes the form of one recipe a week, or writing honestly about what it's like to be a farmer, or maybe it's a funny newsletter that they will look forward to reading. Maybe it's physical changes, like spreading wood chips or gravel on muddy paths. Or adding a portajohn, or better signs.

I used to hate that the goal of capitalism is to constantly grow. What was wrong with producing just enough? A plant grows and then dies (or the leaves fall, or it goes dormant), and then grows again the next year, right? I have come to realize, however, that this view is too simple. An annual plant grows and dies and grows again each year, but depending on changes to the soil, it finds it easier or harder to grow. A *perennial* plant grows and grows and grows. Nothing in life is just cyclical, just doing enough and no more. Everything is also either improving or declining. That is just a natural fact. Soils improve or decline. Wood rots. Species evolve. Stars form and then, billions of years later, they collapse.

There is nothing inherently wrong with growth. My guess is if you have taken the time to read this far in a book that gets this nerdy about business, then you agree. But there are many ways to manifest growth, and the trick is to determine the one that is right for you. It is not always about getting *bigger.* Often it is about getting smarter, commanding a higher price, improving what you have, building a reputation, and leveraging what you know to grow income on the side. But every year, I expect and *plan* to do better than the year before. And then I execute that plan, and react and adjust and make a new plan based on what worked and where we ended up. I don't try to get it perfect right out of the gate. Instead, if I want to do something I just start doing it, and trust that so long as I am thoughtful and observant, and don't create a definitive win/lose situation for myself, I will eventually win.

CHAPTER 7

New Skills

Every farm, no matter the type, requires some specific skills that most of us have no opportunity to learn until we start, whether it's breeding plants, butchering animals, fixing old machinery, driving a tractor, shearing sheep, or understanding soil fertility. These are skills that not everyone has, especially those of us who didn't grow up on a farm. Rather, they are something you stumble across, get pulled in by, fall down the rabbit hole after, and pursue almost despite yourself. Leveraging these skills is an important part of creating a living from your land, but first you must acquire the skill through luck and sweat. For me, the two most obvious examples have been using a scythe and carving wooden spoons.

Scything

Among the wedding gifts my wife and I received, along with various other tools and kitchen implements, was a scythe. It came with a canoe-shaped whetstone and nothing else: no instructions, no information, and no additional repair equipment. I took that at face value and started using it to thrash down some crabgrass on the farm we were working on. Straightaway, I punctured one of the tires on a nearby cart holding a reel of garden hose. After that I was much more cautious.

For the first few years, about once a year I would get out the scythe and batter down some tall weeds, then hang it back up, my back aching. Not once did it occur to me that I could do any research to see how I could improve; I just assumed that it was like using a shovel, that the act of using it would improve my technique, and that any changes would reveal themselves to me along the way.

Boy was I wrong.

It wasn't until I stumbled upon a handle-less scythe blade at a tag sale that I finally started deliberately learning. We were broke enough at the time that I couldn't stomach the idea of buying a new handle for $70–$90, but in my initial poking around online about scythes, I found the website of Peter Vido, the man almost single-handedly responsible for the introduction of European-style scythes into the American market, and the subsequent resurgence of scything that has occurred over the last fifteen years or so. Peter grew up in Eastern Europe, where there remains a culture of using scythes in situations where Americans would use a lawn mower, weed whacker, or brush hog. There is also a tradition there of making your own handles, which Peter took to the next logical step on his homestead in New Brunswick, making handles out of unformed saplings by using naturally occurring curves to achieve the correct shape.

Seeing Peter's scythe handles put into my head that I could make my own handle for this tag sale blade. By this time we were living on the tree farm, and I had access to as many saplings as I could possibly need. Surely there was one out there that was the right shape?

As with anything new and unpracticed, my first sapling handle was pretty bad, my second one was worse, and my third was tolerable. My fourth handle lasted me a year before it finally succumbed to overzealous use. By now I'm on my sixth handle for this particular blade (which has become my main scythe), I have made several dozen others, and my understanding of what makes for a good handle improves with each one.

My understanding of how to scythe similarly improved as I read everything that Peter generously shared on his website (www. scytheconnection.com) and spent additional hours charting and understanding every other person in the United States that has any role in promoting the use of scythes. There are now five good places in the United States to buy a scythe and one in Canada, and unlike my initial scythe, they all offer a full complement of auxiliary gear to keep scythes sharp and properly repaired. Do you really need all the extra gear? Imagine buying a chain saw without the tools to sharpen the chain, or adjust the tension, or even without any bar and chain oil. Yep, you sure do need the extra gear.

Why Scythe?

Before getting into the particulars of how to maintain a scythe, or any more about my saga of figuring it all out, let's fast-forward to where I am now to give a better sense of why the scythe is such a valuable tool on a farm. This is why I kept going with scything, even though I wasn't immediately having success: I could tell, just from watching a three-minute clip of one of Peter's daughters expertly scything a lawn, that this was the tool I had been waiting for.

I had used my share of weed whackers, or string trimmers, or whatever you want to call them. First in my job on the horticultural crew at college, then on the vegetable farm. There are always places that the lawn mower can't reach. I had used my share of power mowers, too, mostly simple push ones but also an extremely brief stint learning to use one of those riding models that can turn on a dime. One of the

Any length grass can be cut with a scythe. Long, short, it doesn't care. This allows for a more flexible management of lawns and meadows. In this photo I've just taken the first pass with the blade which you can see as a crescent of shorn grass in front of me. *Photo by Melissa Patterson.*

original lawn mowers we used when we first moved into our apartment at the tree farm was a Frankenstein's monster cobbled together from the wreckage of its predecessors (which Al never threw out but instead carefully collected in the tobacco barn). This one featured flattened tin cans bolted over sections of housing that had rusted away, and whenever a wheel fell off or the blade got destroyed on a rock, we would rummage around in the graveyard of old lawn mowers until we found something from the last fifty years that could work.

My frustration with lawn mowers wasn't so much having to repair them or even the act of using them. I'm enough of a neat and tidy guy that I love a freshly mown lawn, particularly if the trimmings have been bagged up so you don't get them all over your feet. What I disliked was the all-or-nothing nature of them, how they left a nice lawn most places but when they couldn't cut something at the edge of a tree or building or rock, there was no in between. That grass just stayed uncut unless you were also willing to go around with a trimmer. With a scythed property, everything is a little shaggier but nothing gets so neglected, and the result usually looks more pulled together. I also hated how lawn mowers choked up on grass if it got too tall. This tinged the chore of mowing the lawn with a bit of panic, especially in July when the grass grows like crazy and there is so much else you'd rather be doing than trudging around in circles. Because it's loud, you can't mow early in the morning when it's cool, but instead can only mow when the sun is beating down on you. Because it can't handle tall stuff, you decide in the spring what you want to mow, and anything that grows taller than that is a pain in the neck to cut if you change your mind.

Scything turns all of this on its head. It is a flexible system, because the scythe doesn't care if the grass is ½ inch long or 2 feet long. It will cut it either way. It can cut right up next to things and in tight awkward spaces that the lawn mower won't reach. While it won't get things quite as neat as the lawn mower, it will get everywhere. And you can mow the lawn early in the morning, when it's cool and pleasant, without bothering anybody. In fact, it works better that way. With the exact same scythe, I can mow my lawn, trim around fences and trees and gardens, and mow my small bit of pasture. The grass that is cut can be used to mulch the garden or bed down my chickens. It is a resource, unlike the chopped-up, anaerobic sludge produced by a lawn mower.

Choosing the Right Scythe for You

The superiority of European- versus American-style scythes and traditions is the subject of fierce debate on online forums, where adherents to both argue in great detail about weight, cutting power, metal hardness, and technique. For the purposes of this book, assume that everything I say from here on forward is about European-style scythes, since that is what I use. The reason for my preference boils down to the fact that there is better access to a range of blades and supporting equipment for the European style than for the American style. While many people can find an American-style scythe in their grandfather's shed, at a tag sale, or on eBay, these tools are almost never sharp, undamaged, properly adjusted, and in good working order. I am often approached by people who want to know if their scythe is worth fixing up, and the answer is almost always *no*. If it was your only option, maybe I'd feel differently, but these days there are a number of good options for getting a European setup.

People also often ask me what scythe they should buy, and the answer (as with so many things) is *it depends*. What do you want to do with the scythe? What sort of material are you asking it to cut? What is your own level of skill and energy? Scythes come in a wide range of lengths and shapes. Some are quite short and stout, designed to slash through vines and small brush, much like a machete, hovering just above the ground. Others are slender and medium-sized, more appropriate for trimming close to trees and fences. If you want to use your scythe primarily to clear under an electric fence or trim around trees in an orchard that mostly gets cut by machine, this would be a good choice. Longer blades are good for all-around capability, as they cover the broad area of a lawn or meadow more efficiently than a shorter blade. There are also *extremely* long blades, but these are probably overkill unless you're competing in speed competitions, and are too unwieldy for mixed use. My main blade is 75 cm (30 inches) long, and it does most of what I need. I also keep a slightly shorter scythe for when I need to do a lot of trimming, and a brush scythe for clearing trails and thickets in the Christmas trees. Most people don't need to handle this array of situations, and can get away with one blade, most often in the 65 to 75 cm range (because European blades are *made in Europe*, get used to thinking about them in the metric system).

Scythe blades can be repaired indefinitely as long as you attend to them right away, before the damage grows from the repeated stress of use. Repair is simple, just filing away the damaged bit and re-peening the section of edge.

As I dove deeper into the world of scything, I learned that I was missing a great deal of crucial information and equipment. My original scythe (which I learned was the awkward fusion of a European-style blade on an American-style handle) came with a whetstone and nothing else. As it turns out, a whetstone alone would only keep my blade in peak shape, assuming I knew how to use it (which I didn't), for about four hours. Every couple of hours of use, a European-style blade needs to be *peened*, which is a fancy way of saying it needs to be smooshed thin again between a hammer and a small, portable anvil. After the same length of time, American blades need to have a fresh edge ground onto them using one of those big round whetstones that you crank by hand, or with a metal file. Scythes require continuous maintenance; the idea that all you ever need to do is hone the edge is laughable. Honing is an imprecise sharpening method, which means that over time the edge, while still sharp, will grow fatter and fatter, eventually rendering sharpness irrelevant. In addition, no matter how careful you are (and I was not, at first, naturally careful) the blade will always get dents and nicks, and over time, without the means or knowledge to repair them, those will get worse and might very easily ruin a blade.

There is too much specific, detailed information about scything to include in this book, but luckily it is all easily found on Peter Vido's website. No amount of written information, however, can help you cultivate the mindset you need to exchange a fuel-based tool (weed whacker, lawn mower, tractor) for a human-based tool.

Scything Mindset

Scything has taught me patience. When I started scything, I was used to powering through tasks, damaging both my tools and my body, and making do in the moment until I could finish. This approach does not work with a scythe, as I found out one day when I was scything an embankment where some cut stems of multiflora rose were sticking up; swinging too hard, I hit one with so much force that I put a little tear in the blade. If I had stopped then and fixed it, filed out the tear and re-peened, it would have been fine. But I was close to finishing the section I was trying to mow and I just wanted to get it done, so I continued, and sure enough, when I hit the next rose stub, it caught

Using a handful of grass to clean the scythe blade before honing. Wrap the grass over the spine of the blade, and keep your fingers away from the edge. *Photo by Melissa Patterson.*

Honing is the most dangerous part of scything. Be extra careful and thoughtful about the motion of your hands, or you *will* cut yourself. *Photo by Melissa Patterson.*

It is best to start peening with a jig until you're familiar with the process and know the feel of the edge you are trying to create. A simple peening bench like this one is easy to make, or you can mount the jig in a stump and sit behind it. *Photo by Joshua Klein.*

in the previous crack. All the force of my swing went right into that damaged spot, and I ripped the blade almost in half.

This has not happened since. I am now careful to stop and repair this sort of minor damage when it happens. Similarly, I also stop if I'm getting too tired and catch myself trying to compensate for poor technique or a blade that is just a bit too dull (and needs peening) by swinging extra hard. When you are doing something dumb, and that little voice in your head whispers that it's going to come back to bite you, that voice is always right.

I've also learned to pay attention to where my work falls within the blade's peening cycle. When you peen a scythe blade, it gets squished quite thin, and is at its most delicate. Conversely, every time you stop and hone the blade, you make it just a tiny bit fatter and fatter, which makes it harder to cut grass, but at the same time more robust and better suited to cut woody-stemmed weeds. So for every cutting session between peenings, I try to start by mowing lawn with the freshly peened blade. When that's done I mow tall meadow grasses; by then

Don't forget that I am a lefty. The only change you need to make to these pictures is to switch which hand is doing what. The same is true for photos of me using any tool, except for the scythe itself, which I use right-handed. *Photo by Cecilia Van Driesche.*

the blade has a fat enough edge that it can cut the tougher mix of plants in ditches and embankments with less risk of being damaged. Knowing where you are on this continuum and mowing the appropriate terrain can help you make the most of your time and energy and preserve your blade from abuse.

I almost never peen my brush blade that I use to clear the paths in the grove. It has a much heavier construction, and I want the edge stout enough that it can power through multiflora rose, willow stems, blackberry canes, grapevines, and anything else I want to slash. I cut bigger thickets of these species with loppers, but the brush scythe allows me to bust through a lot of it more quickly. I use my main scythe blade to mow about an acre of meadow each year, which gives me access with my truck to certain areas of the trees, and I also use it cut my lawn at home. I mow my lawn with a scythe partly because I prefer it to using a lawn mower, partly as a means to practice and improve my technique more than I otherwise would, and partly as a promotional stunt. I used to offer my services mowing other people's properties and I still teach people how to use a scythe, and the pictures I post to social media of me mowing my lawn have led to a good bit of the business.

After several years of carving my own handles, I was approached by other people asking if I could make one for them using the same technique of utilizing the natural curves of saplings to achieve the correct shape. Really good scythe handles are expensive to manufacture and difficult to get right, and I realized I could offer people a bespoke handle built to fit the particular size and shape of their body and their scythe blade for less than it would cost them to buy one of these really good handles. Because my handle would be built to fit them perfectly, it would be as good as, if not better than, the manufactured handles; the only catch was I couldn't make it without the person standing in front of me with the blade they wanted fitted, so my handles have necessarily been a limited offering. Still, I make between five and ten handles a year for customers, and as these same people often sign up for a lesson at the same time, the whole package brings in several thousand dollars and provides an economical option to people who would otherwise pay much more for the handle and shipping.

Scything on a Farm

Scything changed the way I manage the farm, giving me an efficient way to cut all the field edges, paths, road edges, embankments, and truck access roads. I don't cut enough for a tractor to be a good investment, and I don't have a place at the farm to keep such equipment. The same is true of a lawn mower, nor would I want to have to mow these places multiple times a summer when I can come in with the scythe and cut them once, in early fall. If I didn't know how to scythe, my best option would probably be to hire a landscaping crew or someone with a brush hog to mow it for me each year, which would cost hundreds of dollars, possibly much more depending on how much I asked them to do. The scythe allows me to manage these areas myself.

Scything also changed how I pruned the Christmas trees themselves. I started off pruning the trees the way Al always had, with a pole pruner. The problem with a pole pruner is that it can only cut one branch at a time, and you need to hold it in one hand while you pull the cord with the other. This makes it slow and exhausting work, even in the best of circumstances and even after building up the required muscles. Most conventional Christmas tree farms shear their trees with long knives, a process that doesn't work for us because our trees often start growing off the stumps at chest height, and the ground is steep and rocky enough to make the use of a stepladder impossible. When brainstorming possible improvements, I got to thinking: What if I could somehow attach a shearing knife to a pole? And then I remembered the scythe blade I ripped in half that had been languishing at the back of the shed because I couldn't bring myself to throw it out.

I harvested a fairly thin, whippy sapling, bolted the blade tip (which is just over a foot long) to the end, and there I was. Now I can use both hands to hold the pole, and because I can cut multiple stems at a time, pruning goes much faster; in one year, the time it took me to prune the grove dropped from eighty hours to forty. Each year since, I have experimented with pruning less and less (Can I just prune the top couple of branches? What if I just cut back the top leader? What if I only cut the top leader if it really needs it?), and while I still think it's important to do some pruning, I've dropped the time to about twenty hours needed to prune the grove. Many trees just don't need to be trimmed. Those that do often just need the top leader

I used the images of mowing jobs to promote both my services and to gain teaching and scythe-building opportunities. This photo I took when scything a giant blueberry patch became the most helpful image of my scything and only took a couple of seconds to take. Every opportunity leads to the next if you take the time to capture it and use it thoughtfully.

It is important to rake up and remove all of the cut hay after mowing; otherwise the next year (or next time if a lawn) the scythe will have to cut through the growing grass *and* this decaying thatch, making it more difficult. Here is ½ acre of hay, which turned into four enormous truckloads that we brought home to bed the chickens through the winter.

cut back and maybe the next whorl of branches down. Overpruning, I realized, was setting my trees back a year or more in the name of a more perfect shape.

––––––––––

Learning to scythe was a big leap forward for the operational ability of my farm. I've described it in some detail not only because I believe scythes are a valuable and underutilized tool for small farms, but also because they offer an example of how a new skill can lead you to reassess the trajectory of your business. For me, becoming serious about scything was the moment I realized that I didn't need to keep doing things the same way Al had, and it was also when I realized that I didn't need to buy a tractor, but could instead replace that enormous cost with my time spread out over the year. Scything gave me options where I thought I had none. New skills are all about options.

Beginning Spoon Carving

Another new skill I developed on the farm has given me options on a far greater scale, eventually becoming the biggest part of my professional identity, what I do with most of my time, and what I am known for around the world. I am talking, oddly enough, about spoon carving. Spoon carving is one of those strange activities that seems so niche that you couldn't possibly make any money doing it. On the face of it, that would be correct. However, by leveraging my life and journey through Instagram I have been able to build a business that currently accounts for a quarter of my yearly income, separate from the farm revenue. By next year it will exceed revenue from the farm. Spoon carving allows me to add value to the deciduous trees on the farm, turning them into a resource that I can sustainably harvest indefinitely.

I started carving wooden spatulas during the fall when my daughter was a year and a half old. It was something I could do while keeping a watchful eye on her as she toddled around the yard. I scoured the winter firewood stacked on our porch for straight-grained pieces, split them open, and carved something quickly. My designs were simple because they had to be. I never knew how much time I had before I needed to change a diaper or tend to a banged-up knee. The wood was dry and hard, my knife was not the greatest, and I leaned heavily on sandpaper to cover up my lack of skill.

The spatulas sold, however, first at the fall festival in town and later at the grove during the tree season. I sold them for $10 each, not a great hourly rate but not awful either, and the money reinforced my desire to do something concretely productive during the mornings or afternoons I had with her. I didn't carve the entire time, but if I made just one thing I could point to and tell myself I had contributed something concrete to the day, something creative, it alleviated some of the exhaustion that comes from being the parent of very young children.

When I look back at those spatulas now I cringe because I knew so little about carving, but I also feel pride, because I recognize that I treated carving like part of the farm business right from the start. Over the years, my skill as a carver has increased dramatically, but that same attitude prevails, as does the desire to make objects that are both functional and beautiful.

Carving wooden spoons has become an important part of my yearly income, but it was not an easy or fast road to this place. *Photo by Meghan Hoagland.*

One of the biggest improvements in my carving skills came when I stopped using sandpaper. In the evenings, after rocking the baby to sleep, I rummaged around online for information about carving, and I discovered an entire online community of spoon carvers from around the world, including the website and blog of Robin Wood, one of the founders of the UK spoon-carving festival Spoonfest. Reading Robin's post on why he stopped sanding inspired me to take what felt like a scary leap. As he put it, when he first started carving he used sandpaper (as I did) "as a substitute for skill and since I knew the tool cuts were going to be removed by sanding there was little incentive to struggle on, getting those cuts as clean as I could. The day I really started to learn was when I threw the sandpaper away."[*] He went on to say that while sanded spoons may feel silky smooth at first—smoother than spoons with a knife finish—after washing, they will feel rougher while the knife-finished spoons will get smoother and smoother with use.

The reason the sanded spoon feels smoother at first is because sawdust fills all the microscopic scratches left by the sandpaper. When the spoon is washed, however, the sawdust washes away and the abraded wood fibers swell, making the surface rough. The only way around this is to wet the finished spoon and sand it while wet, preferably working up through several grits. Phew! That's a lot of sanding. I always hated sanding, so I was immediately inspired to try my hand at achieving a smooth knife finish. This turned out to be much harder than I anticipated.

For starters, I didn't have the right knife. The best knives for carving are called sloyd knives, and are distinguished by a wide, flat bevel that extends all the way to the very edge of the blade, and by their lack of the last, tiny bevel common to almost every other kind of knife. This wide bevel makes it very easy to control the blade and to almost subconsciously know exactly where the blade will cut. The bevel also makes the knife much easier to sharpen to the exceedingly fine degree required to carve comfortably (although I carved for a full year before finally learning how to sharpen well enough to maintain a blade). The most common sloyd knife in the world is made by Morakniv in

---

[*] Robin Wood, "Why I Don't Use Sandpaper," *Robin Wood*, April 22, 2014, http://www.robin-wood.co.uk/wood-craft-blog/2014/04/22/don't-use-sandpaper.

Sweden, and this company has the distinction of creating a carving knife that is both very affordable ($20–$25, depending on the seller) and also the industry standard, used by every professional carver at least some of the time. I had bought a "Mora," but in my ignorance, I bought the wrong blade, a Mora classic, which was shaped exactly like what you would see if you asked the average person to draw a knife blade. Two parallel lines, one of which swoops up to meet the other. It took me another six months of carving for it to dawn on me that what I needed (and what I saw everyone else on the internet using) was the Mora 106. The 106 is meant for carving curvy, delicate things. Its blade is shaped like a long skinny triangle, leaning back on the spine with the cutting edge just slightly curved throughout. This long, tapered tip allows for much smoother cutting through concave curves, where the fatter blade of the Mora classic would catch and chatter, skipping across the wood instead of cutting cleanly. Mora also sells a shorter carving blade, the 120, but the 106 is better at making long, sustained cuts, which translates to smoother surfaces. A year and a half after starting to carve in earnest, I finally coughed up the money for one and was immediately impressed.

I also had the wrong axe. Much of the rough work of spoon carving is done with an axe or hatchet, which can quickly remove a great deal of the excess wood around what will become the final spoon, saving enormous amounts of time and effort. All the carvers online seemed to use very pricey hatchets made by fancy axe companies or craftsmen blacksmiths, but I couldn't justify spending the money, so I scoured tag sales for small hatchets. I used several that were too large and clumsy and one that was way too small before I stumbled upon a mid-sized hatchet with a faceted handle at a tag sale for five bucks. Since then, Robin Wood has started producing some affordably priced axes specifically designed for carving that are quite good; for a time, one of these became my main axe. About a year ago, however, a customer sent me one of those pricey axes I didn't buy in the first place, out of the blue. I was stunned by his generosity and can now say that yes, fancy axes can be totally worth the money.

I originally axed out spoons on the same stump on our porch that I used to split kindling, but after my back almost gave out I finally wised up and put three legs on it, raising it up high enough that I didn't have

My current axe, a Gränsfors Bruk Swedish carving axe. *Photo by Joshua Klein.*

to stoop. I hung my hatchets high on the wall of the porch, out of the reach of the girls, and felt like I was really making progress.

Sharpening

In retrospect, my wife was very tolerant of what might have been easier to humor as a hobby, one that gradually morphed into selling a few spoons here and there after years of improvement. Spending so much time making what in hindsight were pretty ugly spatulas, convinced that someone would buy them, must have seemed crazy to her. It was only the fact that someone *did* buy them that gave the pursuit a hint of legitimacy. But from the start I was focused on selling, production, and pricing.

My interest in carving spoons coincided with my growing interest in making my own scythe handles, and the knife and axe skills I was learning helped me in both pursuits. I don't think I would have done the research that pushed my carving to improve if I hadn't been rummaging around online to learn about scything, and I certainly wouldn't be making scythe handles of the quality I make today if I hadn't developed skills from carving spoons, including how to sharpen my tools properly.

For a long time, whenever I bought a new knife I would think that my carving skill had improved dramatically, only to realize, as the tool slowly dulled with use, that it wasn't me but the knife. At the time, I employed a slurry of different sharpening techniques gleaned from the internet and my own history with sharpening various tools from sailors' knives to loppers, and while I could sharpen a knife at a basic level, I wasn't achieving the degree of sharpness you need to carve comfortably, let alone achieve a smooth, glossy knife finish. Moras were cheap enough that—I am embarrassed to admit—I bought several over the first couple years, rationalizing that spending the $20 was better than spending an hour sharpening and still not getting the blade as sharp as a new knife.

I finally came across another series of videos by Robin Wood, in which he sharpened his knives using sandpaper attached to a block of wood. This was my kind of technique: inexpensive, effective, fast, and portable. Robin used sticky-backed strips of sandpaper stuck to a chunk of MDF (medium-density fiberboard, a type of compressed composite wood manufactured to have a perfectly flat surface) that was held up off a workbench with a big block of wood. Sticky-backed sandpaper proved difficult to find and I didn't have any MDF kicking about, but after watching the videos several times I decided to see if just wrapping a loose sheet of sandpaper around a smooth block of wood was effective, and to my delight it was. It has taken some refining since then, but I have found using automotive-grade sandpaper (which is available in the necessary fine grits) wrapped around one of my daughters' building blocks to be an easy, cheap, and portable system. Most importantly, however, I am now able to achieve a greater level of sharpness for my knives than they come with from the factory. While I have since begun exploring ceramic stones and strops, I would recommend that anyone starting out use sandpaper to sharpen. The best place to see my current knife-sharpening techniques in action is to visit my YouTube channel (just search for my name) and watch one of the videos where I sharpen my tools.

Taking the Leap

For several years I derived income from the Christmas tree farm, the scientific editing business with my father, and a seasonal job managing a property for a Massachusetts-based organization that bought and protected significant properties. The latter job was part trail building, part barn repair, part garden design and maintenance, and part teaching workshops. Cecilia and I used to joke that these three jobs were my stool of employment: Kick out any one of the legs and the other two were not enough to stand on, but all three together were just barely enough.

Over the course of three years, my summer job evolved into more of a mow-the-lawn-and-shut-up situation as the organization changed leadership and direction. Thankfully, the growth of my two businesses made me feel empowered enough to leave, so I left. My plan was to get a smartphone, join Instagram, and start selling wooden spoons full-time. At that point I had been carving for about two years, and I saw enough happening online to make me think that it was possible to sell enough spoons to replace the income from that second job. To put this in perspective, I was the only person I knew at the time who didn't have a cell phone. In fact, with the exception of a one-year stint with a flip phone ten years earlier, I had *never* had a cell phone, and swore I would never need one. But I also recognized that Instagram was where the wooden spoon movement was happening, and my desire to succeed and to take full control of my income was greater than my desire to stay on my high horse, so I did an about-face.

I thought that by posting on Instagram my talent would quickly be recognized, I would be able to find fancy stores in cities that would want some of my work, and I'd be able to build up a nice little customer base. I thought it would be relatively easy and fast. It was neither.

I also thought I was pretty good at spoon carving. I was not.

I spent that first winter sending out over 150 spoons to over seventy stores, trying to drum up business. Only two stores bit, but neither account turned into anything. The repetition of that work, however, made me a much better carver, and also made me appreciate the fact that perhaps working with retail stores wasn't the right sales avenue. I spent the following summer working for a local farm a couple of days a week to make ends meet, and kept carving and posting my work to Instagram.

Marketing on Instagram (or any social media platform) is an odd gambit. Half of success comes from the sheer quality of the work (both what you're marketing and the picture you take of it), and the other half comes from how many followers you have and your connection with them. I started out with zero followers, knew nothing about photography, and didn't understand how hashtags worked. My goal was to post at least one picture a day, however, and the discipline of this commitment kept me both carving and improving as a photographer. Thankfully, today's phone cameras do a lot for you, so the photography skills required to take a decent picture with your phone have much more to do with anticipating the moment, framing, and choosing the appropriate lighting (which usually just means turning off all the lights if taking a picture during the day). I spent my evenings scouring Instagram for makers whose work I aspired to and whose client base I wanted, and I wrote down all the hashtags they used. By the end of that first summer, eight months in, I was a much better photographer and had built up enough of a following that I was starting to sell my work. Around this time, I had a crisis of faith around my pricing.

When I first started selling my work at the grove, before I was on Instagram and knew what everyone else was doing and charging, I sold spoons and spatulas for $10. I knew it wasn't much, but I also recognized that they would likely be an unplanned impulse purchase from my tree customers and also that I was a total beginner. The second Christmas, I sold spoons for $15. Then I saw that people I was trying to emulate were selling their spoons for $25–$40, and I thought I should do that, too. That decision almost sank me.

I felt in my gut that $25 was too much for what I was producing, and I could sense it in the reactions of other people, too. Friends, family, neighbors, people on Instagram, prospective wholesale customers— all expressed surprise at what I was charging. I defended myself by pointing to the examples of other carvers on Instagram. Needless to say, I didn't sell very many spoons, and I spent a lot of time feeling conflicted and guilty both for asking that much and for taking the money of the few people who did buy one. My skills were developing at such a rate that I would look back at three-month-old work and realize that it wasn't very good. So how could I justify charging all that money for something I would eclipse in a matter of months?

Finally, just before the Christmas season, I decided to cut my prices in half, and that's when things really took off. I started selling every spoon I could make, giving me even more incentive to carve, and accelerating the pace of my improvement. The surest sign that I was on the right track, however, was that I became proud both of what I was making and of the value I was offering. I was no longer embarrassed to describe my carving business to my neighbors, and my focus shifted from achieving recognition to recognizing how I could bring more value to my followers and customers.

Coming out of that Christmas season, my goal was to carve some-thing every day. I posted it on Instagram and it usually sold in about five minutes. Gradually, I began getting commissions, first a spoon here or there, and then more and more, until by June, I had a six-week waiting list, which has grown to two months as of writing this (a year later), and would be longer except that I have quadrupled the amount of work I schedule for each day. I now book at least $120 of work for each day, which can take me anywhere from four to six hours, depending on what I'm making. Just before last Christmas I increased my prices by $3 almost across the board, and this October I will increase them further. In a year and a half I will be back at the prices I tried to start at, except my work will be much, much better and I will be operating with a several-month waiting period.

As my waiting list has grown, I have pushed to increase the amount I carve in a day.

About half of the spoon-carving business at this point is selling spoon blanks to aspiring carvers who live in cities or regions without access to quality wood.

I also began teaching spoon carving and scything, occasionally at other venues but most often out of my home, and my schedule has filled over the last year to the point where I deliberately limit the amount to four or five lessons or workshops a month so I don't burn out. As with selling spoons, teaching started off slowly. In order to start teaching I had to get extra professional and liability insurance, which makes it less of a profitable use of my time than just carving, but allows me to help people improve their carving and fits into my long-term strategy, which is to raise my profile within the spoon-carving scene as a teacher and resource.

Both carving and teaching are examples of leveraged income, taking a skill you gained on the farm (or an opportunity you have because of the farm) and expanding it. Anything can become leveraged income, even if it's not directly related to farming. You might prune and graft fruit trees, peddle Popsicles, make award-winning hot sauces, or cut and deliver firewood with draft horses (all examples off the top of my head from local farmers I know). A product or service can be a good candidate for leveraging into income if it's something that is extra valuable to others, is something you have in abundance, or is something you are passionate about (or any combination thereof). Depending on how much demand there is for what you want to do, you will need to act strategically in order to grow it into a significant part of your income.

While carving wooden spoons may seem to some like a diversion from my farm and the work it entails, I would argue that it is actually a significant contribution. Not only have sales of spoons doubled or tripled in each of the last four years, the spoons take something I have in abundance—extra wood—and turn it into a source of income. By using the farm as a springboard to learn and promote this skill, I am now at a place where I carve full-time, year-round, and have a long waiting list. I teach private lessons at my home, with enough demand that I actually have to ration out how many I teach. Three years ago I made $1,000 from spoons. Two years ago I made $3,000. Last year, sales from spoons and lessons exceeded $10,000, and this year I'm aiming for $15,000–$20,000. Carving represents only a portion of our yearly income, but it is an example of how you can leverage a skill and build an independent living based on many small streams of income. The trick is that it takes years to reach a place where your skill can start earning a significant amount of money, not only because you need to learn it well enough to produce something of quality and to understand your market, but also because it just takes time to build a reputation, particularly if the opportunity for growth is in a sector you have previously ignored, such as online sales.

————————

Both scything and carving are pursuits that initially seemed frivolous to many, but I cared about them deeply and was determined to make them work. It has taken years of getting good enough to charge for my skill and to build up enough of a customer base that I now make 30 percent of my yearly income from these two ventures, separate from the seasonal farm income. These are no longer insignificant amounts but represent real increases in our quality of life.

Whatever you want to do as a side hustle, get good at it. Really good. The stranger it is, or the more competition there is (opposite ends of the spectrum, I know), the better you will need to be to justify the value of what you are doing or to stand out from the crowd. Give yourself the time (years, if necessary) to become skilled enough and to build a reputation and customer base that will support you at a meaningful level. If you can, use economic incentives to drive your growth. Part of my strategy in slashing my spoon prices was to create as much incentive as possible for me to continue carving. Even when I started needing a waiting list, I decided to keep my prices low, embracing the flow of work and pushing myself to carve more than I otherwise would have. I would rather carve five things and make $100 than carve one thing and make $100, because in the long term, carving five things better serves my development as a spoon carver.

Don't assume that you have to limit yourself to being a maker, teacher, or service provider. Do all three. In my case, I make scythe handles for customers. I also teach scything, and people used to hire me to scythe their properties before I phased that out for lack of time. All three fed off one another, increasing my expertise and bolstering my reputation. Similarly, I carve spoons, teach carving, and also sell spoon blanks to fellow carvers who have less access to wood than I do. Each of these parts brings in thousands of dollars of income, and each supports the other.

The handles on these sloyd knives were inspired by the shape I was making for the scythe grips. All my knives are currently made by Matt White of Temple Mountain Woodcraft.

Embrace social media. You may be doing so already, but most land-based small businesses (my own included) could do a better job promoting themselves on social media, even if their customer bases are local, and even when they think they don't need to. If your primary base is hyper local, Facebook might be the right choice for communicating with them. Are you more of a maker (or a farmer, for that matter)? Use Instagram. Twitter is preferable when communicating ideas rather than images, so if photography isn't your thing but you have a lot to say about food policy and land use reform, that could be the spot where you gain traction. Whatever app is the best fit, social media is a big part of how you establish a reputation in today's marketplace. Even if people can find you in other ways, they will use social media to learn more about you, and you can use it in turn to establish connections with them. The key to social media is to be disciplined about it; an inactive account is worse than no account at all. Figure out what sort of posting frequency works for you (daily? weekly?) and make it happen. If you are trying to build a business, more is better, since between the ever-changing nature of social media sites' algorithms, and the sheer amount of content out there, posts become obsolete fast; the more content you put out, the more eyeballs will see it. You might be tempted to think that the quality of your work will naturally draw in customers. And that may be true. But if you aren't content with your current sales, or worry about having all of your eggs in the one basket of local sales, then work on your social media skills.

The most important paradigm shift you can make in any business is to start thinking about how you can provide the most value to your customers rather than what you can do to succeed yourself, and that is also true for side hustles. I spent the first couple of years thinking about spoon carving only in terms of what I liked to do. I refused to carve certain shapes (such as coffee scoops), only to realize, when I finally agreed to make some, that I actually love carving them. Time and again, I have found that when a customer asks if I can make something for them that I have never considered, I should say yes. For instance, I never thought to offer spoon blanks (a billet of wood roughly shaped with an axe and ready to carve) to fellow spoon carvers, but when someone asked I said yes. Now, over half of my orders are for blanks, and I'm on target to earn over $10,000 from these alone. The service of making spoon blanks has led me to serve and connect with many more customers than would have bought my spoons if that was all I was offering; now they buy spoons, travel great distances for lessons, and recommend me to their friends. I have also gotten *much* better at making blanks than I was a year ago.

I'm a big believer in trying stuff out. We can never really know what will work, what will take off, or what we actually like doing until we try it. Figuring out how to dabble in something without spending a lot of money or effort is crucial for sorting through all the different ways you could be spending your time and energy. For instance, when I first

Spoon carving started as a productive thing I could do to fill in the gaps of my schedule. Five years in, it has become the central focus of my time for most of the year. *Photo by Meghan Hoagland.*

tossed around the idea of trying to make some money from scything, Cecilia suggested I make some flyers with a Sharpie and computer paper. How delightfully old school! I made three copies, posted them at the local watering holes, and that summer got three jobs scything people's properties. The internet makes it easy and tempting to drop money on flyers and business cards that you end up realizing two weeks later are not what you need. Starting simple, with a marker and a sheet of paper, can help you refine your wording, your angle, and your prices. This is the promotional equivalent of impermanent infrastructure.

I'm still figuring *everything* out, and I expect I will be for the rest of my life. Life is one reinvention after another, or evolution if you prefer to think of it that way. However you want to frame it, the truth is we *never* get it right straight out of the gate, so don't agonize about it. Just start trying stuff. Follow what works and what feels right, and figure out a pragmatic balance of the two. Leveraging income takes time. I think the main reason businesses of all sizes fail is because people vastly underestimate how much time it takes to build a reputation. Overestimation of one's skill can be another reason, but anyone you see hitting a stride that you want to emulate has likely put in far more time and effort than you appreciate. Building a customer base can be slow going. But this doesn't mean that it's not happening or that the venture is destined for failure, as long as you're prepared to be in it for the long haul.

Scything and spoon carving are what has worked for me, based on my interests, what my land provides, and the opportunities and customers in my region. They may or may not make sense as income streams from your land depending on your own skills, interests, and opportunities. Al Pieropan grew strawberries to sell and an entire peach orchard from pits (he taught me that while peaches don't breed true, about half of the seedlings turn into good fruit-bearing trees). He used a chain saw sawmill to cut big logs into slabs, which he stored in his barns and turned into beautiful rustic benches. As James Rebanks wrote in his excellent book *The Shepherd's Life*, "the geographic constraints of the farm are permanent, but within them, we are always looking for an angle." Give your angle—or four or five or six angles—time. Let them develop naturally, but also be ready to push them along.

If you spend enough time and effort leveraging your skills off the farm, you might find yourself making more money there than from your land's core opportunities. I would argue for a number of reasons that it is important that you keep farming, though, even if you find it to be the least profitable use of your time. The first reason is that your success will always, at least in part, stem from your farming experience. Keep farming and you will keep learning. If you have been farming for five years, think just how much more you will understand after ten. If you have been farming for ten, imagine twenty. Another reason to continue is that farming wonderfully balances most other forms of work. Even when it is hard, and sweaty, and buggy, or cold and icy, I find in it an immediacy and necessity that I find hardly anywhere else. There is a satisfaction that comes from tending to a piece of land, improving it year after year. Don't give that up to chase dollars, but rather consider it a creative constraint. Figure out how to work around it, and you might end up in a place you never thought you'd be. I certainly have. If you told me five years ago that I would be carving spoons and teaching carving and scything full-time, I would not have believed it possible. No one believed it was possible. And yet here we are, exploring new waters, all made possible by a strange little Christmas tree farm we stumbled into.

The last, and most important, reason to keep farming is that *the world needs all the farmers it can get.* Chances are you already feel this way or you wouldn't be reading this book.

CHAPTER 8

Telling Your Story

I f you take over someone's farm, as we did, to one extent or another you also inherit their treatment of the land, their reputation, and sometimes even their name. You represent the next chapter in the story of that place. If the previous owners did a great job and you're just taking the baton for the next leg of the journey, then your branding strategy (as long as you keep up the quality) might be straightforward. If you are taking over a failing farm business, you might have some work to do to rehabilitate its image or relationships. If you start your own farm or land-based business, you have the opportunity to write the first chapter of its story, and that's not without its challenges either; in fact, it can be overwhelming. The good news is that in the age of social media, it is easier than ever to define yourself. Unless you're a commodity farmer or selling only to wholesalers, like it or not, land-based enterprises require storytelling, and you need to take seriously the job of using all the elements of your farm's identity—name, physical space, reputation, online presence, and real-world behavior—to support that identity.

Once Cecilia and I realized we needed to continue calling our farm what everyone in town was already calling it, the Pieropan Christmas Tree Farm, the rest of the brand identity we wanted to create flowed from that name. We worked with a graphic designer friend of ours to create the initial logo, font, and website. While it is easy enough to build your own website these days, it's worth working with a graphic

Al Pieropan only ever had one sign with the farm name on it.

The original sketch for the logo, inspired by Al's logo.

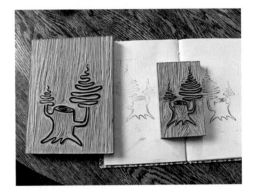

The sketch turned into linoleum cut prints.

designer if you are not clear on exactly what you want. Our instructions to Seth were to use a font that evoked Art Deco, with a bit of 1950s-era diner thrown in, a mix we both loved in its own right but also felt would help build on the nostalgia people have around our farm and around Christmas more generally. We came up with a logo that took the one Al used (a straight line for the trunk of the tree with a squiggly line denoting the branches) and updated it to show the stump bearing two trees, one little and one big. The new logo emphasizes our growing method, which is what sets us apart from other tree farms, while at

Our own sign where Al's used to hang. Note the way the colors echo those of the road sign and the trees.

the same time giving a nod to Al's legacy. We first made some pencil sketches of this idea, then a number of drawings with marker until one of them was clearly better than the rest. I gridded out this drawing and transferred it to a linoleum block, then cut away everything but the lines of the drawing. I sent this linoleum cut to Seth, who digitized the print the block made and then cleaned up some of the lines to make the image less cluttered.

For colors, Cecilia and I wanted something different from the classic white-lettering-on-dark-green-background motif that is the standard choice of all Christmas tree farms. We liked the idea of using naturally occurring color combinations found in the balsam groves (something we had done with our tiny house, painting the door pale green with a scarlet doorknob in homage to a species of lichen that is pale green with a scarlet tip). We went to the paint store and got a couple of different shades of green, some dark and some light. I fell in love with a pale, bluish green that reminded me of the color of blue spruce, but Cecilia quite rightly pointed out that it felt cold. The best color combination juxtaposed the dark green of branches with the much brighter, almost yellow color of the new growth each spring; after painting test sticks to see which color we preferred for either background or lettering, we ultimately settled on a dark green background with yellow-green lettering.

While this color combination does not stand out much from late spring through the summer when the leaves are on the trees, by the end of November, and especially when there is snow on the ground, these colors make our signs really pop out. If we were selling a different main product at a different time of year we would likely have chosen a different color combination, and it is important to consider not just how your color combination looks in isolation on a computer screen or business card, but also how it looks against its likely backdrop in the larger world. Will most of your customers see it on a sign in front of a green pasture? On a business card at the bottom of a cardboard box? On a grocery store shelf, competing for attention with a dozen other brands? As a logo on a plastic bag?

Fonts are also important toward conveying the story you want to tell the world about what you do. Seth selected a handful of fonts that felt similar to the main font on our signage to create a website and flyer that were both evocative of the past and modern at the same time. In particular, the font he chose for our name perfectly evokes the feeling of early twentieth century that I wanted to pay homage to. It is eye catching, immediately identifiable as us, and easy to lay out and paint freehand. The fonts Seth chose for the secondary text work similarly, and just as well. It would have been easy to assume that you should choose one font and stick with it, but Seth showed us that the right mix of fonts can be much more effective at communicating discrete pieces of information while at the same time building a textured world for the identity than any one font alone. Choosing a strong mix of fonts can be tricky, but start paying attention and you will see them everywhere, particularly on well-made websites. Take note of combinations you like and write down what you like about them, or the font names if you can figure that out.

Seth built our first website using these fonts and our general color palette of greens and white, and for a number of years it was our main way of presenting ourselves to the world. After four years the information began to feel dated (although the look and feel did not) and so we replaced it with a less well-designed Wordpress site that we can more easily update ourselves. Ironically, this website now feels dated and needs an overhaul, which will happen before this book comes out.

Identity Basics

The basic components of your visual brand are: name, logo, font, and color scheme. All four of these should be rooted in the emotion you want to evoke in your customers. Before choosing anything, ask yourself what assumptions you want someone to make about your business. Do you want them to think of you as traditional? Modern? Classic? Do you want them to feel invigorated by your identity? Challenged? Comforted?

Many names are forgettable. Many names are not easily searchable online, an important consideration these days. Sometimes a name just makes the most sense (like in our situation) because the farm you're taking over has a history that you want to honor. Other times you are forging something new. Choosing a name that can grow and encompass a changing farm identity can be important (not something we did), but so can choosing a name that is highly specific so that people remember what you do. A good name combines unexpected words to surprise and delight. A good name suggests a logo. Some of my favorite local farm names are Queen's Greens, Atlas Farm (with a fantastic logo of Atlas bearing up under a load of produce instead of the world), and Town Farm (so called because it is in town). Remember that customers who visit your farm (if your farm allows visitors) will often leave out the word *farm* when referring to it. So when we say that we're going to the Atlas Farm Store, we just say that we're going to Atlas. Make sure your name works under those circumstances.

The lantern color, the shingle siding, and even the type of hook the lantern hangs from are all carefully chosen to evoke a certain feeling and support a certain color palette.

When colors, materials, fonts, and textures come together to support a consistent tone, you create a unique and memorable experience for the customer.

Logos can be challenging to conceive and create. Sometimes you don't even need one. In this age of smartphones with their tiny icons, many logos are too small, too detailed, too spidery. They are easy to ignore. It's best, especially if your business will do a lot of online business, to either choose something extremely simple and geometric or hire an artist, graphic designer, or illustrator to distill the essence of your idea using as few lines as possible. When a friend and I recently started a project on Instagram called @spoonesaurus, I asked a cousin of mine who is a cartoonist to create an avatar for the account, of a *T. rex* carving a wooden spoon with a goofy grin on its face. I also asked her to use bright colors. She captured the absurdity I was hoping for—the big dinosaur using its teeny tiny little arms—so wonderfully that I have been asked numerous times if we intend to make T-shirts. That is the power of a good logo. People want to wear it.

Font and logo often work together hand in hand, so that the name of the business is always written in that font, to the point where the words would look funny in any other font, and the arrangement of the logo also feels sacrosanct. With our farm, for example, we almost always position our logo to the upper right of the name.

For me, the simpler the color scheme the better. Choose two colors, or occasionally set the darker of those two colors against white if you need to change it up. Make sure the lettering contrasts enough to really pop out from the background. Unexpected colors for a farm are a good bet: pinks and oranges and yellows and pale blues and greens—there's a reason neon signs are eye catching. Often, using two tones of the same color but different shades (like our dark green and light green) is a lovely choice. Eyes are sensitive to contrast, so if both of your colors have the same degree of darkness (like primary colors, for instance) then they won't stand out as distinct from one another. But navy blue with pale blue lettering? Heavenly.

Consistency

Ultimately you want your name, logo, color, and font to work together in such a way that people associate the font and color scheme with your brand, recognize your logo even from a distance, and easily remember your name when they think of your product. You can achieve this by being consistent with your presentation. Every time you display your farm name, use the font, colors, and if possible the logo. On our farm, the best example of utilizing this combination is our road signs.

Because our farm is in the middle of nowhere (I used to be quite proud of the fact that from our house, you could drive through two intersections in either direction without encountering a stop sign), road signs are crucial for helping people find our place during the Christmas season. We post signs at every intersection out to a radius of 4 miles (6.4 km), all featuring our name, logo, and a large arrow pointing people in the appropriate direction. Over the years I realized that these signs needed to be bigger and bolder than I had at first imagined, particularly the ones that weren't at intersections where people would be coming to a stop. In order to effectively communicate something to a person driving 40 miles an hour (64 kph), a sign (and its lettering and arrow) needs to be about four times larger than you are probably envisioning right now. It is also important that the font, logo, and coloring be consistent across all of your signage, because those physical cues create a gestalt that is identifiable from much farther away.

We have twelve such seasonal road signs. When possible, they are bolted with carriage bolts and wingnuts to existing signposts, taking

Almost all of our signs employ the same background and font colors, and the lettering is consistent among our website, signs, and literature.

advantage of the holes drilled along the length of these posts. Where a street sign isn't available, we screw signs into trees or telephone poles or mount them on stakes. We also have two giant signs that stay up year-round that identify our farm and the You-Cut grove in particular, five signs directing You-Cut customers away from our wholesale groves and down to the You-Cut grove, and three signs posted at the grove that feature a legal disclaimer required by law to protect us from someone suing if they get hurt. We have signs for wreath pricing, signs that give directions for paying if we aren't around, signs that we take to festivals and markets, and signs that stay at our house where we sell trees and wreaths off our front fence. On all of these, the coloring and font are consistent. One of my proudest moments of the past year was when a man came up to me at the grove and said that he was a marketing consultant and that our farm was the best example he had ever seen of signs being obviously handmade but also beautifully done and consistent.

The handmade aspect is important. While it is certainly possible to buy banners and print out signs for a reasonable price these days, there is something simple and authentic about a hand-painted sign that a slicker alternative could never achieve. I strive for a crisper presentation than the classic spray paint on a piece of plywood approach,

A hand-painted sign where just the right amount of care is taken with the lettering has an energy to it that no printed sign can emulate.

Painting Signs by Hand

I am not a professional sign painter, but here are the tricks I've learned over the years from having painted several dozen signs with a wide range of sizes and content. While I started out using a stencil that I cut out of cardboard to pencil in the first batch of road signs, I quickly became comfortable enough with the font and spacing to do everything freehand, and I would recommend anyone do the same as quickly as possible. Stencils are time-consuming to use, and can prevent you from making the signs you need on the fly because the process seems too long to be worth it.

First, determine the size of the sign you need and what it's going to say or convey. A good rule of thumb is to make it four times bigger than you initially think it should be. As much as possible, strip the wording down to the bare minimum, even if it's in a situation where you think people will stop and read it. No one reads a long sign.

If you use plywood, knock down the edges a bit with coarse sandpaper, as paint won't stick to a super-sharp edge. Paint the background color at least two coats on each side, and hit the edge grain each time, making for a total of four coats on the edges. This will help the signs last a lot longer, as the most common failure of plywood signs is delamination from water getting into the edge. I have also found it invaluable over the years to keep some blank sign bases kicking around, so I can whip up a quick sign as needed without painting the base first.

Make a small mockup on a sheet of paper at roughly the same proportions, to get a sense of where the lines of text should fall on the sign. Then draw these lines in pencil on the sign as faintly as you can. This is another reason to avoid white signs with dark lettering, as these pencil lines will be invisible on a dark background but will be apparent at close quarters on a white sign. Also, white signs look great at first but get grungy quickly.

Practice writing the sign's wording in your chosen font, both small and to scale. Over time you will understand the ins and outs of your font and won't need to do this every time, but for a while at least, this practice will keep you from making a mistake that stands out like a sore thumb. Where are the lines thick? Where are they thin? If you

Our hand-painted OPEN sign. *Photo by Meghan Hoagland.*

haven't studied typography, you may be surprised how many details go into creating a distinct and recognizable font. Practice how each letter is formed until you can close your eyes and envision it accurately.

Letter in your sign, starting with the middle words on each line and working out to the edges to help ensure they are centered. Again, your paper mockup is your guide to determine where the words fall. Pencil in not just the bones of the letters but also where they are thick and where they are thin. Pencil in any serifs, the small lines that some fonts have at the ends of each stroke, if appropriate. A word on fonts: The more minimal the font, the trickier it is to paint, as their simplicity makes every wiggle and inaccuracy stand out. On the other hand, fonts with serifs and other flourishes can be a pain in the neck. Choosing a font that has enough complexity to mask imperfection but is easy to paint is an important consideration if you anticipate needing more than a handful of signs.

If you are right-handed, paint the sign as you would write normally, from left to right. If you are left-handed, as I am, paint from right to left, so that your hand can rest on the sign without smearing any paint.

Use a thinner brush than you think you will need. It is much easier to use a tiny brush and make the thickness of the lines exactly what you want them to be than it is to use a wider brush, at least at first. It would take a lot of practice to get better results with a big brush than what you can achieve the first time around with a smaller brush.

but I also want to keep my signs looking unpretentious. I take most of the signs down each year and store them under cover, but even so some of them are starting to look beat up. Rather than touch them up each year, I embrace this weathered aesthetic. When I paint new signs, I pencil in the letters quickly so that they are relatively evenly spaced but don't look fussed over. My painting style has also gotten looser as I have gained a better mastery of my chosen font, where lines should be thick and where they should be thin.

Own Yourself

If you're leveraging farm income into side businesses, you may need to create a brand identity separate from your farm name, as I did when I started carving wooden spoons year-round. I briefly tried another business name that could encompass not only spoon carving but anything else I wanted to do, but it was unsuccessful. When I eventually paused to analyze branding within the spoon-carving field, I realized something. All the top spoon carvers are known by their names: Barn, Jojo, Robin, Yoav, Jogge, Fritiof, Anja. When a carver uses a separate business name (and many of them don't), it usually goes completely unnoticed. Furthermore, there appears to be a sub-conscious distinction between how we think about businesses versus craftspeople. If you go by a business name, the baseline assumption is that your primary identity is commerce; if you are known by your own name, on the other hand, you are much more likely to be seen as a teacher or independent artisan, someone who is in some way defining the conversation about the craft, not just selling it.

Now, I have always shied away from using my name professionally. Emmet Van Driesche is a mouthful, usually misspelled, often mis-pronounced, and generally impossible to remember accurately. But I knew I wanted to teach and to have a voice in the philosophical canon of the field, not just carve spoons. Thankfully, the internet has made the accurate propagation of my name much easier, as people link to things and can take advantage of autofill in search engines to find me even when they can't remember how to spell my name. So rather than come up with another generic-sounding business name, I decided to bite the bullet and put myself out there as just me.

One of my daily spoon photos. A coffee scoop carved from birch.

I didn't have my epiphany about business names versus personal names until about a year in, so for a while my Instagram handle (or name) was @pieropantrees, because our farm website is www .pieropantrees.com. Meanwhile, I was trying to use the name Anchor Goods as an umbrella for all the things I wanted to make and sell: not only spoons, but also hand-bound notebooks, leather and canvas bags, things I had been making for myself for years. It was an awkward fit. People were confused by the name and it didn't tell them instantly what I did. Although my farm sells almost entirely locally, I was trying to promote Anchor Goods both in a global space and at local farmers markets that already knew me as the Pieropan Christmas Tree Farm, which made things even more confusing.

After a year of struggling to gain traction as Anchor Goods, I had the epiphany, dropped it altogether, and changed my Instagram handle to @emmet_van_driesche. The effects were immediate. While my follower count was already gathering momentum and did not experience a spike, people's sense of *who I was* finally clicked. Sales increased. Interaction increased. I felt like I wasn't hiding behind a company name anymore, and as a result both my photographs and their posts became much more personal. Not inappropriately so, but I started portraying myself honestly as a person, and I found to my astonishment (although I really shouldn't have been surprised) that people appreciated that much more than the "professional" image I had been trying to project before. My posts started telling the story of my journey as a farmer and craftsman. I had stopped faking it till I made it and was instead sharing my path. That turned out to be a much more powerful paradigm.

Having a smartphone and taking pictures every day for my Instagram account has been my single most important business development in the last two years. The direct business it generates aside, the discipline of taking photos every day has meant that I have become a much better photographer, and I now have a large trove of photos that I can use on my website. Instagram's chronological portfolio of photographs is ideal for showcasing handmade tools and crafts, and like Twitter, it also effectively employs hashtags for greater searchability. Because the platform is image-based, captions can seem like an afterthought, but they shouldn't be. Most users go for short and

Taking Photographs

I am not a professional photographer, nor have I done any training or studying of it other than taking lots of photographs for the last couple of years. For the first three years, I used my smartphone exclusively, because it's always in my pocket, allowing me to capture moments that would otherwise pass unnoticed. These tips are, accordingly, just notes about composition and process based on my experience of what has worked for me. These are intended to be general principles, not just how to photograph a product in a way that grabs the eye, but also how to capture moments.

I have recently started using a Sony a-6000 with a simple manual lens, and would recommend that once you get your feet under you with the smartphone you take the leap. The ability to achieve a tight depth of field can sometimes be duplicated by in-phone apps, but few phone cameras can generate the number of pixels of a real camera, and the more pixels you have, the more options you have for how you use the image.

Less is usually more. The number one thing that makes a photograph memorable for me is a stripped-down composition and color palette. This can be achieved by zooming or getting closer, cropping, or choosing a frame that naturally has only a little going on. If in doubt, frame in closer.

Don't be afraid of color. Wonderful color is everywhere. Often it is crowded out to the point where we don't appreciate it. Use the techniques above to focus in on one color and let it be felt. A photo where someone's amazing red sweater fills the frame and all you see is that and their hands is more striking than a shot of them in the red sweater against a parking lot of cars. The more you can strip away, the better what is left stands out.

Light is key. While phone camera technology is getting better and better at handling low light conditions, harsh midday light is still tricky to navigate. The time of day dictates what I take pictures of, and the key is to tailor what I'm photographing to what the conditions of the moment will allow. If there is a picture that can only be captured at a certain time of day, see if you can remember to take it on an overcast day. Early morning also makes for good pictures, which works well for farmers or anyone working outdoors. Artificial light never looks as good as natural light, so if you are shooting indoors, turn off any lights or lamps, even if you don't think will make any difference.

Lead the eye. Farms are full of wonderful geometry, from the rows of crops to the geometric spacing of seedlings to the arches of greenhouses. Use some naturally occurring structure or geometry to format the picture for a stronger impression.

Use the rule of thirds. If possible, frame landscape-style photographs such that the focus takes up one-third of the space, with the rest of the background taking up the remaining two-thirds, or by lining up the horizon at an angle across the shot. For some reason, our eyes like to see things in threes, including spacing.

Embrace imperfection. Luckily for farmers, imperfections look great in photographs, whether it's weathered wood siding, rust on a tool, or a kale plant succumbing to winter. Use weathered surfaces as a background to make your produce or other products look more perfect by comparison.

Anticipate the moment. With action shots, the time to have your camera out and ready to go is the instant *before* something awesome

snarky, while others go for long and earnest; I tend to fall somewhere in between but definitely hit those extremes. Hashtags can also add commentary themselves, either funny, more to be laughed at than to be followed as a link to similar pictures, or earnestly, used to add your picture to a body of other work that you feel matches your aesthetic or philosophy.

I use hashtags primarily as a way to be found by other people. Social media is a numbers game; the more followers you have, the easier it is for more people to discover you. One of the most important ways to grow your base of followers when starting from zero (as I did) is to use happens, not as it occurs. It will take some time to develop the knack for knowing when to get ready. Taking too many photos can ruin the moment and inhibit the natural action of others. But quietly observing some human encounter, pulling out the phone, and discreetly snapping one picture? Often totally worth it. I find that my first one or two pictures of something are usually the best, and if I keep taking pictures after this point, they become increasingly contrived and stiff. Do your best and then move on.

Keep it real. Using photography as a means of noticing what *is* rather than contriving a scene usually results in better photos than I would otherwise take. The discipline of leaving things as they are is a creative constraint that has led to some of my best pictures. If I'm photographing an object, like a spoon, I put it down in the middle of some scenario or surface, and take the picture. This brings to the picture all the energy of a real situation that would immediately be lost if I brought in props. Cropping, choosing the light setting, and framing in thirds are all important, but this rule keeps me open to some strange experiments that often work quite well. It also breathes life into pictures by giving little glimpses and clues about the life around me, what one of my daughters is reading or what we will have for dinner. These real moments have a power that a contrived moment never could.

multiple hashtags on every single post. That way, anyone browsing the pictures for that hashtag will be able to see your post, link back to your feed, and decide to follow you if they like what they see. Pay attention to how your feed comes across to someone checking it out for the first time. Are all the photos good? Do you have something to say? I regularly go back and edit my feed, removing the weakest photographs to improve the quality of this first impression.

About six months after this change, I created a website with my name for the domain (emmetvandriesche.com), something I thought only writers and politicians did. Because my identity as an individual was easily understood, as opposed to a business name, it could cover any pursuit I wanted to undertake, including those not directly related to the farm. Now people can find out about me, buy spoons, and arrange for lessons without needing the backstory of the Pieropan Christmas Tree Farm. This isn't to say that the farm isn't important, or that its identity doesn't carry meaning, just that it is equally important to know when to separate some of what you do from the farm name. If you're a farm, and also producing goods that are not strictly agricultural (like the spoons and scythe handles I make), then it can be an awkward fit to sell them under your farm name.

However, your farm identity is still important, especially when you're starting out, because it gives you a base of expertise from which to speak, teach, or create. The farm can lend you initial legitimacy, before your name has any of its own. But at a certain point, you may find it appropriate to reclaim your own name as the source of your authority.

Reputation

Identity is inextricable from reputation. This has always been true and it is only more so now, with the internet's tendency to make reviews, comments, and articles linger. Make sure that your daily actions match the identity you want to have. Be kind to your customers. Be kind to everybody. Give them the unexpected deal. Go the extra mile to help them. Show up when you say you're going to show up. Keep to your deadlines. It is tempting to think of your identity as just what you put out into the world, but in truth your identity is the sum total of everyone's experiences with you and your farm. You never know which one of those voices will be amplified, so the best course (and the right course) is to do the best you can every time.

For me, this means walking up into the grove with a lantern in the dark to help a lady who arrived super late from Boston select a tree. It means inviting countless kids and elders into my hut to thaw out by the woodstove on bitterly cold days. It means harvesting balsam

Running a farm or business involves balancing a lot of details. Maintaining a sense of humor is crucial under all circumstances, because you will inevitably drop things. *Photo by Meghan Hoagland.*

greens in the snow and ice even when it's miserable out because I gave someone my word I'd have them ready. It means always being honest with everyone. It means giving someone a price break from the usual flat rate for a tiny tree, without them asking.

Being consistent about putting the customer first earns moral capital that helps perpetuate itself. For example, last year I totally dropped the ball and forgot about an order for some bales of balsam. Not only was the woman's order not there when her husband showed up, but I had left for the day. She called me, understandably disappointed and upset. But although I couldn't help her that night, and her husband had already driven back home, I was able to arrange for another customer, a florist who was picking up some bales that evening, to grab some from a third party's order that wasn't going to be picked up until the next day. The florist lived near enough to the woman that she could easily go pick them up from him. The florist is not a big customer of mine, but I've known him for years and we have a good relationship. He was happy to help me in my time of need because I had built up the moral capital over the years, not through some extraordinary act, but just through charging fair prices, keeping my word (as best I can, obviously, as this story points out), and being respectful. In this case, instead of telling the woman that she was out of luck, I apologized and then solved the problem that I had created.

Maintaining your reputation as a seller online can be trickier. For instance, a fellow spoon carver on Instagram recently accused me of stealing his idea; instead of confronting me directly, his approach was to call me out publicly, again and again. Two things helped me weather this situation: how I responded in the moment, and my existing reputation. I responded as respectfully as I could, reaching out to him privately to explain how I was not stealing his idea, but that what I had carved was a customer request. I also acknowledged the fact that I carved and posted a photo of a spoon very similar to his and not long after his post, but since I hadn't stolen anything I stood by my work. When my private communication didn't get anywhere, I stated my position publicly in my own feed, on his feed, and in the several large discussions about plagiarism that built up on the feeds of several other colleagues. In all of these I was respectful when he was not, even though I was upset.

To my surprise, most people came to my defense, both privately and publicly, so much so that he deleted his posts with the original accusations. He then demanded that I similarly remove all the posts I had made discussing the situation. This I refused, on the grounds that since it was my reputation that had been smeared, I wanted my defense to remain public.

My reputation prior to the incident is what inspired people to reach out to me with their support, and my calmness and respectful attitude during it helped to cement that reputation. Ironically, the whole thing was a positive, though painful, experience for me. I found myself thrust into the center of an argument about ideas and ownership, design and plagiarism, sharing and reputation. On the one side were a small group of professional carvers who were more established and had created an atmosphere of territoriality, where they claimed certain design elements as their own and jealously guarded against anyone else using them. On the other side were the masses of amateur carvers and younger professionals like myself who viewed these design elements as common wealth, basic shapes that were fair game. Largely because of being accused by this fellow spoon carver so publicly, I found myself championing the ideals of sharing, respectful discourse, and equal rights to designs in a way that I would never have sought out otherwise. You might think this sort of ordeal will never happen to you, that you don't need to plan for it, but you never know what is coming around the bend. Live as though this type of conflict is inevitable, and you will be in a much better position if it happens.

It is inaccurate to say that we can "choose the identity" of our business, because we define and redefine both ourselves and our businesses every day, with every act. It is the accumulated form of these actions that creates an identity. Identity is the opinion of your neighbor and the image someone has in their head halfway around the world. It is your reputation with your customers and your reputation online. It is your own name and your farm's name. It is the way you portray yourself online and the reality on the ground. It is the sum of the relationships you build and the grand idea in your head. It is forever a work in progress.

The ability to tell your story is as important as any other tool for your land-based business.

Commonsense Reminders for Gaining and Maintaining Your Reputation

Reputation is not static. It can change in an instant, and yet it is also the sum of your words and deeds. While it might seem trivial to articulate some of these basic things that build a reputation, in this day and age we can probably never read or hear or talk about these things enough. Your farm's success and your success as a human will lean heavily on your reputation.

- Treat everyone with respect. Even rude people. Even people deliberately trying to harm you. Respect is the foundation of all peaceful relationships.
- Keep your word. If you say you are going to show up, show up. If you say you will deliver something, deliver it on time. It is always better to underpromise and overdeliver, so don't make promises you can't keep.
- Tell the truth. Sometimes this means saying something when it needs to be said. Sometimes this means acknowledging your own mistakes or wrongdoing.
- Be a champion. Use your unique advantages to be a voice for those who are voiceless. Stand up for what you believe.
- The right thing to do is always the right thing to do.

Whenever you start something, whether a farm or some other business, it is easy to feel paralyzed about how you want to portray yourself, what you want to be called, what you want to *do*, even. The great thing about thinking of identity as simply telling your story is that it takes the pressure off. You don't need to have it all figured out before you begin. You don't need to feel stuck with the name you have chosen, what your farm is already known for, or how you have positioned yourself. Just tell your story. Change things up. Toss stuff at the wall and see what sticks. Articulate these changes and give them context using social media. Telling your story keeps you tuned in to yourself and what you want and need out of life, and it keeps you tuned in to the needs of your land.

Farming is inescapably land-based (even if you think it's not, because you have a concentrated animal-feeding operation or a hydroponics greenhouse, don't fool yourself; all of life is land- or ocean-based), and the story of your land precedes you and will continue after you. Having both of these ends of the time line in mind as you make your choices will help guide you to where you want to be. What is the story of your land and how does that inform your options? What do you want your own legacy on the land to be? What do you want your legacy to be as a person? In the next chapter, I discuss how the answers to these questions guide your future moves.

CHAPTER 9

Future Moves

Every year that we've owned the tree farm has been different: different goals, different readiness, different resources. At first, we reacted to Al's decisions. He passed on more trees to our care, and we scrambled to make use of them and drum up more business. Now we're in a place where we need to respond more to the grove, what it will give and what it needs. Our goal isn't always to make more money each year, but is to position ourselves in a stronger place for the future. Sometimes the changes are practical in the short term, while many others are geared toward long, slow shifts that will span years.

I believe that life is change. Even if all you want is to keep things the same, you need to work to keep it that way. If what you want is something more than what you have, that requires even more work, and even more planning. And because the world is changing around us, in most cases that means that we *must* change over the years, or start sinking below the rising costs of goods and living, a changing work environment, changing tax pressures on land, changing expectations around doing business, changing needs as we and our children grow older. We are only as strong as our next move.

So what is that next move? Ideally, it's something that flows from what you are already doing. It's the next logical step in growth, or the hole in the market that you are uniquely positioned to see and fill. It's the dream you've always had, or some practical thing that you do because you need to provide for your family. Whatever it is, you should start planning for it, and promoting it, now.

Farmers, by necessity, are hustlers. We know that if we do something differently next year and the weather cooperates, it could work out. That extra plowed acre, or bunch of calves, or whatever, could mean the difference between a fat year and a lean year. The problem is that the yearly growing cycle encourages planning only at that scale, which by some standards is a long time, but is too short when it comes to the sort of meaningful, long-term shifts that are what make a farm thrive over a lifetime. You can lull yourself into not doing enough for the land, year after year, each time thinking that you'll do better next year, enact that crop rotation next year, fix the leaking roof next year, remineralize the fields next year. To make the long-term choices that are right for your farm as a business and right for your land, you need a long-term plan, the discipline to carry it through, and the ability to shift the details in reaction to changes in the market and on the ground. In short, you need to do what farmers have always done: You need to adapt.

Grove Rehabilitation

As I mentioned earlier in this book, over the last few years I've been experimenting with pruning less, and have gotten a better sense of how little I can get away with without the trees getting too leggy. I plan to spend most of the time I've saved from lighter pruning by pushing back deciduous trees and cleaning up stumps that have become overgrown, as demand is starting to outstrip our supply. Anything I can do to open up more stumps to productive growth will be a good investment three years from now. By pruning trees more lightly, I will also keep more of them at the target harvest size this year; if pruned heavily, a tree that looks ready will end up looking too small.

My other goal is to keep destroying multiflora rose thickets. In the last two years we have plowed through a lot of the worst ones, but now I need to hold the line to keep them from coming back, and keep pushing to eliminate the rest. My goal is to have the farm (or at least the part we lease) free of serious multiflora rose in four years. This year will be the year of mopping up a bunch of smaller thickets before

The grove is full of stumps that need to be cut back to select the best young trees and give them the space to grow as swiftly and evenly as possible. *Photo by Meghan Hoagland.*

Sometimes quite drastic amounts need to be removed. When possible this is combined with harvesting greens so that these branches can be sold, but it is worth clearing out stumps at any time of year. *Photo by Meghan Hoagland.*

The final step of rehabilitating a stump is to select the new growth with the strongest leaders and remove the rest. *Photo by Meghan Hoagland.*

they expand, and then the next several years will be spent tackling the really huge thickets that remain. After about three years of keeping the rose from resprouting, which I do by simply cutting any new shoots, I should be able to plant new balsam in these areas and not have to worry about them getting swallowed up. With invasive species, you need to have a plan for the land going forward. It is not enough to simply eradicate the rose. It will always come back if you don't create the opportunity for something else to thrive in its place.

The third prong of grove rehabilitation is to address the yellowing that occurs in so many areas due to wet conditions. Two years ago I tested nitrogen addition, to no effect, so this year I will test iron supplementation, and also push to harvest these areas first, before they have a chance to turn color. As I wrote in chapter 6, if iron does not have the desired effect, I will need to assess whether it is worth trying to dig drainage ditches of some sort to help lower the water table in these areas.

All three of these moves are long-term strategic actions that will continue in years to come. While I might see some benefit this year in the form of paths that are easier to navigate, stumps that produce better-colored greens, and a lack of rose brambles to tear up my clothes, their real fruit will be born several years from now, in a farm that is better kept, more productive, and more profitable.

Leaning Up

This coming year will also mark a shift in how we approach the growth of our business. For the last several years, we have increasingly employed seasonal workers, accepting as many orders as possible, and pushing for maximum production. We have now reached the end of this path as a smart option, largely because our demand is starting to outstrip what the land can sustainably produce. This year, we will lean up our operation to increase profits while reducing the load on the grove. While not having employees removes many costs associated with hiring (most notably Social Security along with Workers' Compensation and Liability Insurance), the real heart of this strategy is saying no to orders. For this to work, we need to have a clear idea of the relative profitability of different things we do and make. Here is the list of the different ways I spend my time, in descending order from most profitable to least: retail trees, wholesale trees, greens, retail wreaths, wholesale wreaths, roping, spoons, teaching. The catch is that, while it is true that selling bulk balsam greens is more profitable in terms of my time than making and selling wreaths, the amount of greens is limited by what the grove will produce, while I can max out the number of wreaths I can make and not over-tap the grove's production of greens.

Our systems are all in place; the adjustments to what we do at this point are just minor tweaks. I use my gut to tell me whether or not I can fill another tree order as the season progresses, but at a certain point I just need to say no. I fill a reasonable number of greens orders, and then as many wreath orders as I think I have time for; I have stopped making roping, in part because it is hard on my body but also because I recognize that it is the least profitable use of my time, and

It is always worth my time to tie wreaths, and this understanding of the relative profitability of the different opportunities I have is the key to making the most of my effort.

if my time is the limiting factor then it would be irresponsible of me to make it. I do almost no spoon carving during the tree season, trying instead to stock up on inventory ahead of time, when I don't have such profitable ways to use my time. For an example of how stark the contrast can be, I typically make about $20 an hour carving spoons. Tying wreaths, I can make $75–$120 an hour, depending on which size I'm making, even factoring in the time it takes to gather the greens. I can harvest $100–$150 an hour of greens when the cutting is good. So every hour spent carving when I could be tying wreaths is money I'm throwing away.

Leaning up the operation means making every process more efficient, and looking for any small area where I can lower expenses, produce more, or expend less effort. Leaning up is not the same as generating more business, and it is not the same as raising prices, although both will have the same overall effect. Leaning up is about capturing the revenue that already exists.

Willow

Planting basket willow is another long-term plan for the farm. Eight years ago I bought one whip of willow from Fedco, thinking I'd start making wicker baskets. Each year for the past eight years, I've planted out the whips produced by that original plant, cutting them back each spring and simply sticking the cuttings into the ground, where almost all successfully take root. When we bought our current house, I planted several rows in a wet swale of the chicken meadow, and another row along the edge of the road where water pools.

A year ago I took a basket-weaving class taught by a young woman who lived nearby, and she remarked how she knew of no one selling white willow (willow with the bark stripped off) in the United States, and how expensive it was to have it shipped from Europe. Seeing a hole in the existing market, I have ramped up taking the year's cuttings from the previously planted willow and planting it out in some of the wettest areas of the grove, where no balsam has ever been planted. I first roughly chopped down the existing shrubby growth and then planted the willow every which way, aiming for roughly one every couple of feet. The goal is to lay the foundation for substantial willow beds that will make use of these wet areas that are currently unproductive. I can then harvest, soak, and strip the willow in the spring when my workload is the lightest and market it to US basket makers at a lower price than that of European white willow. These beds are probably three years away from starting to throw off long rods, so until then my plan is to use the existing growth to expand the plantings. There is a vast amount of land that is currently too wet for balsam that is comprised of native willow, multiflora rose, and whatever else grows up between cuttings that occur every ten years or so. Converting this land into willow beds would allow me to manage these areas profitably while still maintaining a complex ecosystem, with plenty of other plants coexisting with the willow.

Most of the willow planted at my home is pollarded (pruned by removing the upper branches of the tree), with the first cut established several feet off the ground. Largely this is so that I can trim more easily around the trunks with a scythe. I plan on coppicing the willow I planted at the grove, however, cutting it right down to the ground. Coppicing densely packed willow tends to produce more long stems, as they all race for the sunlight from the same starting point. I have also started planting willow along the fence to the chicken run at our house. There I've planted the willow quite closely together, just a couple of inches (5 cm) apart. They will be pollarded 4 or 5 feet (1.2 to 1.5 m) up, and in four years when the trunks are several inches thick, will have a gap of just an inch or two between them; when I tear out the sagging chicken wire fence I'll have a new, living fence. Over time, the trunks might even fuse, forming a solid wall of willow! I'm excited to see what happens, and excited to create a structure that will become more beautiful as it ages, unlike the chicken wire that gets more and more dilapidated. I can see myself as an old man, clipping the chicken fence each spring, and then weaving baskets from the trimmings.

Stripping the willow bark requires a device called a brake, which looks like a metal version of an old fashioned clothespin. The idea is to clamp the brake to some upright post or tree trunk, and pull each willow rod through it, where the bark is lightly crushed and becomes easy to strip away. Before stripping, it is important to soak the willow by standing bundles upright in buckets of water. After a week or so, roots will start to appear, indicating that the bark is loose and ready to

The original willow whip I planted seven years ago. Willow grows tremendously quickly, and within three years can start throwing off significant marketable rods. In this photo you can also see the three main barns at the farmhouse.

be stripped. The rods then get stored under cover and allowed to dry and shrink before resoaking them for use.

Our backyard on the north side of the house is an ideal place to manage these operations. I already have a rough tarp barn that I use to store signs for the farm, wood for carving during the winter, and my chain saw gear. I could easily retrofit it to have a storage rack for willow. In addition, I acquired an old canoe that should work as a soaking tank for preparing willow to weave, or soaking the ends of a number of bundles. When I acquire a brake, I have a stand of small trees that would be perfect for clamping the device, and the whole setup is shaded except during the last couple of hours of the afternoon. Now I just need to have the patience to wait three more years. I can use the intervening time to make connections within the basket-weaving world, so that when I have white willow ready to ship, I will already have customers lined up.

Spoonesaurus Magazine is a magazine by spoon carvers, for spoon carvers.

Spoonesaurus

Perhaps the biggest future move related to the farm (although at this point practically in name only, as I deliberately leveraged it away from the farm as much as possible in case a long-term lease doesn't work out) is where I hope to take my spoon carving. Last year, I started a project on Instagram with my friend Matt White called @spoonesaurus, which originally was intended to showcase short videos describing, in nerdy detail, one particular spoon-carving tip or another. We have since expanded the project by hosting gatherings and starting a magazine.

I decided to start *Spoonesaurus Magazine* in large part because I couldn't find a comparable magazine devoted exclusively to the world of spoon carving. It also felt like a good way to combine my strengths (spoon carving and writing) and leverage my reputation to create a business that could be scaled in a way that actual spoon carving could not. The number of magazine subscribers can triple, while I cannot triple the amount of spoons I carve.

The first issue of *Spoonesaurus* was a proof-of-concept that was funded out of pocket, intended to be given away at spoon-carving festivals and as an add-on to the first true issue to subscribers. These days you can purchase drag-and-drop software that allows you to lay out pictures and text quite easily, which you can then send to the printers where they shuffle the sequence into the correct order for printing with the click of a button. I found this software frustrating to muddle through at first, but after playing around with it for a few hours, I figured out enough to make it do what I wanted. Although I dreaded the layout process for the first issue, I'm looking forward to the next one.

Matt and I also hosted the first Spoonesaurus Gathering at his house in New Hampshire, and a second one at my house. These are basically just free meet-ups of fellow spoon carvers, a chance to hang out and carve together, and it is unlikely that they will become a source of income in the future. But it feels important to do what we can to bring our community together and to provide as many resources as possible, both in person and online. On a practical note, I have also found that the more I give of myself, the more work comes to me. Call it karma, call it reputation, call it friction, the more I share what I know for free, the more orders for spoons and arrangements for lessons I receive.

Starting Spoonesaurus with Matt has made me appreciate how empowering it is to bring something into existence that you wished was already happening. Having a partner, rather than working alone, allowed us to dream big and give ourselves permission to take on roles we would not feel confident enough to step into on our own.

In terms of how much time in my day that I can realistically devote to spoon carving given the number of orders I receive, I'm about at capacity. Each year I raise my prices when I close my books to prepare for Christmas, so my income will grow accordingly over time, but the amount of carving I can do each day is about five to seven spoons (or about sixty spoon blanks), depending on what types of spoons I'm carving and how many other obligations I need to fulfill. The growth of my spoon-carving business comes from mastering my technique, pushing for more efficient processes, and continuing to use social media to build as broad a base of followers as possible. In three years or so, I plan to build a proper workshop, although such a build will be a massive onetime cost to gain these changes. But having a dedicated space for my work will allow me to host lessons without needing to take over the house, and to host gatherings more gracefully. It will give me a heated space to work in the winter (right now I bundle up and work in an unheated greenhouse and in my kitchen). It will provide me with a space truly organized around the business, and not the muddled mixture of work and home I have now.

The goal is for the workshop to be about 16 × 20 feet (5 × 6 m), heated with a woodstove that can use the scraps produced by the spoon-carving process, and perhaps with a supplemental propane heater. The space will be largely open and unstructured, so that it can accommodate large gatherings as easily as one-on-one workshops. It will look out on the house and gardens, and yet be just 50 feet (15 m) from the house doorstep.

While I could continue to work from the greenhouse, woodshed, and kitchen indefinitely, and probably save money that way, the goal is to set myself up for a lifetime of this sort of work. This will be the structure I work in as an old man. Right now my ambitions and ability are limited by the spaces where I work. For example, when I axe out spoon blanks now, I tend to stockpile the bagged-up blanks on a bench in the woodshed. When it's time to box them up for the weekly mail run, I have to haul them upstairs to our bed, the only surface big enough to spread them out so they stay organized. I box them up and then haul them back downstairs to the truck. In the workshop, I plan to have a long bench along one wall, with a shelf below it to stage these blanks as they are made and bagged up (the bagging preserves their moisture). Below that, I plan to store a row of wire crates where I can toss the larger scraps of wood from the carving process so they can dry before being burned. Everything will happen in one place.

I am quite glad, however, that I have not yet built the workshop. Had I been in a position to build it right away, the result would have been overly designed and too small. In the two years I have been dreaming about it, my thinking about the workshop design has refined, and experience has made me realize what I actually need from a space. As much as I love diving into a project right away, I will grudgingly admit that there are real benefits to waiting, pondering, hesitating, planning, sitting on something, and generally taking your time, particularly when it comes to physical structures. The end result (as long as you don't fizzle out) will almost certainly be better than your initial idea. In general, building our tiny house, then the You-Cut hut, and now planning this workshop has made me appreciate the value of a wide-open floor plan and as much square footage as you can afford. Both are worth holding out for.

Scything

Scything, which only accounts for less than $1,000 of income a year, has gotten the short end of the stick in terms of my time and attention. I started the Western Massachusetts Scything Association several years ago as a way to share skills and resources on a local level, but ultimately I found that it was easiest just to handle inquiries personally rather than try to rally the other people I knew who were promoting scythes to work together. Each year I make maybe ten handles for people and teach maybe five to ten lessons. While I used to promote myself as available for scything jobs, I now have enough demand for spoons that the exertion scything requires is no longer worth it.

There is also much less public interest in scything than in farming or spoon carving, making it the most fringe activity I do. However, I firmly believe that the scythe has an important place in small farming practices, and so I see my scything practice as mostly playing a waiting game: doing enough of it, in terms of actual scything and teaching and making scythe handles, to be prepared to ramp it up should momentum start to build. For instance, if a seed and tool company were to embrace the scythe in the next ten years or so as part of an initiative to develop and make available innovative tools for small farmers, it will change how the farming community sees scything, and increase demand for my work.

Writing

I envision writing as an increasingly important part of what I do going forward. Most of this writing will be in the form of my blog, which I've written weekly for the last year now. The blog was an important trial run for many of the ideas in this book, and will continue to be the place where I explore the intersection of my craft and what it means to create a business and career. *Spoonesaurus Magazine* will be an outlet for more technical writing about spoon carving, while I continue to pursue a number of book ideas.

Writing dovetails nicely with farming because it is a good winter activity, when I have a bit more downtime. Many farmers, perhaps most famously Eliot Coleman and Joel Salatin, have written about farming and leveraged a career on their ideas far beyond the reach of their farm as a business. If you are not a writer already, this doesn't mean you don't have it in you. Record yourself talking and then transcribe it, if need be. Don't get bogged down in making it perfect, and keep in mind that an idea actually on the page, captured, is better than ten that come and go without leaving a trace. The blog has been helpful for this, giving me a weekly opportunity to explore something that I otherwise wouldn't articulate. I sit down, write for thirty minutes, and then post it, without agonizing about it and rarely even reading it over. It is an exercise more than a product, but I have gotten a great deal of feedback from people who appreciate the conversational tone and the ideas I explore.

When writing this book, I found it helpful to get up at five o'clock every weekday and write until seven. When you factor in the time to get dressed, walk the puppy, and get settled, we are talking an hour and a half a day. In an hour and a half I can write fifteen hundred words on a good day, and five hundred words on a bad day; most days I fall somewhere in between. But the words stack up, day after day after day. I may not enter that transcendental state in which Arthur Miller wrote Act I of *Death of a Salesman* in a single day, but I can achieve sufficient flow to feel inspired, and frankly I don't have the space in my life right now to work that way. I do try to stop writing *before* I reach the end of what I have to say. Knowing what I want to say next allows me to start back up easily, which can be hard if I've stopped the day before at some natural conclusion.

Website

The tree farm website is due for an overhaul, which will be our third in nine years, on par with the industry standard for turning over websites every three years. It feels surprising a website can be outdated even when the physical infrastructure of the farm has changed so little. But your website is where many people get their first impression of you, where they find essential information and learn your story. Over time, as your story expands and your perspective on your work shifts, you might want to reframe how you describe what you do, or where you place the emphasis on your operations. If nothing else, photographs, particularly of children, get outdated quickly. Our current website was made before I had a smartphone, before I posted lots of shots of the farm on Instagram, before one of our dogs died, before we bought our current home, before I had a book about the farm to promote. It feels, as one customer put it to me this last season, "delightfully retro." Yes, well. That will now change.

Our farm began as a discrete business whose work (both in its schedule and in the public's eye) was associated with a particular time of year. Over the years our identity has shifted, or expanded, and what we do sprawls across the entire year. Some of this growth has come from taking skills and opportunities from the farm and leveraging them into separate income streams. Some has come from writing or teaching. We have expanded the core of what the farm produces, while at the same time keeping our options open by making efficient choices with our infrastructure, tools, and cultural practices. We farm with very low overhead costs, and the balance of income between multiple streams keeps us from putting more pressure on the land than it can sustain.

All of this can happen with any farm, or any piece of land. Start by assessing both what the land can do and what you *want* to do, and then find the overlap between these two things and what the market wants. Remember that you won't always find a market that is local, and you won't always find it already established. You can succeed nevertheless.

Work with natural processes to improve the quality of your land. Rotationally graze pastures using tight paddocks to stimulate better grass growth. Allow slash piles to rot back slowly in woodlands to recycle nutrients and support complex ecosystems. Alternate cover crop mixes between vegetable crops. Pack bed animal manure to capture nutrients from leaching and off-gassing, and then windrow the result to allow it to compost.

Likewise, choose equipment that is appropriate for your scale. You might be able to let go of your tractor if you could get by with a big rototiller and hand tools. Or maybe you are currently using hand tools and it's time to start investing in more machinery so you can better use the amount of land you have. The land base, your own affinities, and what size business fits your life and market potential all go into determining what level of equipment makes sense. In general, if you can start small with hand tools, you will be able to pivot more readily as you figure out what works and what doesn't. I do not use any of the tools today that I started out with eight years ago. Switching to my current set cost me just a couple of hundred dollars. Compare that with the cost and hassle of trading in one tractor for another, or buying a piece of equipment only to realize that you need the next size up or one with a different feature. If you are on a trajectory to grow in size, then you will need to grapple with these distinctions sooner or later. Delay too long and you might be missing out on a certain momentum, both among your customers and in your life; jump too soon and you assume a great deal of debt, or at least expense, before knowing what you need.

The same is true for infrastructure. The time will come, if you are savvy and successful in your business, to pour many thousands of dollars into some infrastructure project, whether it's a greenhouse, barn, milking parlor, certified kitchen, or workshop. But before that time, if you can go through a couple of cheap, temporary permutations of your idea, you will gain tremendously valuable insights into what works, what doesn't, and how you should actually build the permanent solution. Consider committing to a couple of years of research and development with impermanent prototypes to hammer out your idea and build revenue to a point where it makes sense to take the leap.

Don't forget the basics of interacting with the public while you are at it. Make your name, logo, and color scheme memorable and consistent across your signs, literature, website, and business card. Keep your market tables elevated and use a long tablecloth. Remember to smile! Wear clean clothes. Pay attention to the tiniest details of how you present your product. Primp that wreath. Mist those vegetables every half hour. Make sure your labels are straight and easy to read. Be consistent in the quality of your offerings. Don't run out of product.

Set up an administrative regime that you understand and that works for you. Hire a bookkeeper to help you a couple of hours each month or year. Establish a routine for each sale or market. Batch your processes so you can be efficient about keeping the books and going to the bank and the post office. Know what you are legally required to collect in the form of taxes and what laws regulate your business.

Understand that being your own boss means you will work longer and harder than you would work for anyone else. Understand that you are building a reputation with everything you do and that this reputation will be your legacy in ten years, not the best of what you produce or the worst, but the sum total of every interaction, every experience someone has with you and your farm, both in the world and online. Manage all of this deliberately rather than just letting it happen. You are in control of your image through the thoughtful use of social media, your website, the physical space of your farm, and how you behave in every situation.

Be patient! It takes time to build anything. You cannot compare yourself to peers who have been walking the same road as you for years. It will take you years to gain the knowledge and experience they have. It will take years to build your soil, or to finally beat back the invasive species, or to breed the herd of your dreams, or to afford the barn you envision. It will take years to build a reputation, and it will take more years to be worthy of that reputation. It will take years to build the relationships with customers and other businesses that are the backbone of any farm. It will take years and still you will be learning, and doing more, and dreaming bigger, different dreams than the ones you started out with.

Give yourself that time. Life is long and full of changes.

———————

The morning after Thanksgiving is cold. Some years it's warm and sunny, other years rainy; this year it's clear but biting. As I steer the truck through the turn onto the final stretch of dirt road leading to the You-Cut grove, I note with satisfaction that I'm the first one here.

My breath is visible in great clouds that linger in the air as I unlock the hut and immediately busy myself lighting a fire in the woodstove. Everything is at hand: newspaper, some crushed pinecones, and empty wire spools for kindling topped with three small logs. I strike a match and hold it to the edge of the paper until I'm satisfied that the flame has taken. While the hut slowly warms I busy myself with opening up: I flip the sign on the door, slide the OPEN sign into place, and haul a fresh bale of greens from under the tarp barn into position next to the workbench.

Before Thanksgiving I left the wreath wall stocked with wreaths, but two of them have sold, so I fetch the appropriate sizes from the poles of wreaths under the tarp and hang them on the wall while I decorate them. One needs a bow, which I fetch from the box of extras under the workbench, while the other needs a spray of winterberry and a cluster of cones, both of which are stashed behind the hut. The wall fully stocked once more, I walk down the road to one end of the You-Cut grove to make sure the bow saws I staged at the entrance to that side are still hanging on their tree. When I look back, the smoke curling out of the hut chimney tells me that things are warm inside and ready for me to begin.

The first car pulls up as I'm finishing my first wreath. I reach up to the shelf where my phone and speaker are blasting out Mavis Staples and dial it down a few decibels. Through the window I watch the

Ten years ago, just starting out, Cecilia and I walk out into the trees to do some work. *Photo by Peter Reich.*

family get out of the car. I recognize them; they've been coming here for years, part of a regular crowd whose tradition is to get their tree the day after Thanksgiving, no matter the weather. They bustle about pulling on jackets and hats while the mom walks over to the hut. When she gets close I slide open the ice cream window and lean out on my forearms. The Staples Singers back up Mavis, "I'll take you there," and the woman starts grinning as she gets close enough to hear the music.

"Hey!" I say. "Nice to see you again. How was your year?"

ACKNOWLEDGMENTS

First and foremost, I would like to thank Al Pieropan's family for being so supportive of us throughout the years. I would also like to thank Al himself, who passed away in October 2018, for giving us this opportunity and for being committed in life to having his work carry on after him. Without that resolve, this would not have happened.

Thanks also to editor Michael Metivier for seeing my book more clearly than I could myself, and to the entire team at Chelsea Green. As I write this, I have barely begun to interact and work with so many of you, but it has already been a delight and I look forward to getting to know you more as we embark on the journey of getting this book out into the world. It would not be possible without you.

Giant thanks to my family. I am the luckiest guy in the world to have the whole extended circus of you. I love you all so much. Finally, to my wife Cecilia, my best friend, who made this whole thing happen by kicking me out of bed each morning so I would write. Thank you. I love you.

Knots

The Bowline

If you only resolve to learn one knot, this should be it. This is the knot you use to tie a rope or string to any object. Need to tie a bit of paracord to a water bottle? Use a bowline. Need to tie a painter to a canoe? Use a bowline. Its power stems from the fact that no matter how tightly the bowline (pronounced "BO-lin") gets pulled, you can always untie it, because it has a hard curve that you can bend backward to start to loosen it (sailors call this "breaking the back"). Bowlines are usually taught using the old "rabbit comes out of the hole, goes around the tree and back down into the hole" mnemonic, but that is not how I like to teach it. First, a definition: the "standing end" of a rope refers to the end that is attached to something. Imagine a shoelace. The part of the shoelace going into the first lacing hole on the shoe is the standing end. The other end (the bit with the plastic end on the shoelace) is called the "bitter end" by sailors, but here I will just call it the "loose end."

1. Hold the rope with the standing end in one hand and the loose end in the other, so that it forms a U in toward your body (it doesn't matter which hand holds which end: The bowline, like any knot, can be tied either way). The length of this curve roughly corresponds to the size loop you will make, so make it bigger or smaller by just adjusting the size of the U.

2. Take the loose end and place it across the top of the standing end so that it forms a cross, with the actual cut end of the rope overlapping by just an inch or two (2.5 to 5 cm) (figure 1).

3. Place your hand on top of that cross with your fingers in the same orientation as the loose end and twist your hand and the loose end together to form a small loop in the standing end with the loose end come up through that loop. Note that both the standing end and the loose end are still pointing away from you (figure 2).

4. Pull the loose end around behind the standing end and tuck it back down into the small loop (essentially a hairpin turn around the standing end) (figure 3).

5. Tighten the bowline by holding the loose end doubled back on itself in one hand and the standing end in the other hand and pulling in opposite directions.

6. To untie the bowline, flip it over and pry back on the hairpin turn. No matter how tight the bowline, there will always be some give in this area, which can be increased and used to loosen the entire knot.

The Bowline on a Bight

The loop that this knot creates is useful because it can be made without using the rope ends. It gets its name from having a form similar to the bowline when tied, including a back you can break to untie it. I prefer the bowline on a bight over similar knots that are simpler to explain (the butterfly knot, the alpine rider, the Spanish bowline) because this one is easier to locate in a precise place. This makes the knot more useful because its sole purpose is to use in combination with a rolling hitch to gain mechanical advantage when pulling a rope tight. When creating these mechanical advantage sys-

tems, you often need the loop created by this knot to be in a certain place because of the dimensions of the system and the length of rope; this precision is easier to achieve with this knot than with others. The fact that you can untie it is crucial, because this knot always gets pulled tight. So while you *could* replace it with an overhand knot tied into a loop, that knot would be permanent, while this one you can untie.

Unlike many other knots, which I tie ambidextrously, I always tie a bowline on a bight the same way.

1. Form a loop in the middle of a length of rope so that the piece that lies on top comes out of the loop clockwise (figure 1), then continue making one turn around the loop you just made, as if tightening the loop with a belt across its middle, slightly compressing it in the process (figure 2). This forms two smaller loops, one at eleven o'clock and one at five o'clock.

2. Flip the five o'clock loop up and over the eleven o'clock loop and pull the eleven o'clock loop through (figure 2).

3. Tighten by pulling with your left hand to the left and the ring and pinkie finger of your right hand to the right (figure 3), while you adjust the size of the loop with your right forefinger and thumb and use your middle finger to control the five o'clock loop as it tightens down (figure 4). This may sound tricky, but when you actually break it down, this is exactly what each finger does.

4. Untie the knot by breaking the back, the same as for a regular bowline.

The Rolling Hitch

The primary purpose of the rolling hitch is to tie back to itself when using a rope to strap something down, and while you can technically use it to tie a rope to another object, I never use it that way. I tend to use it to pin the rope when it has a lot of tension and goes around something (usually a tree or a post of my truck frame), and then I use a rolling hitch to hold that tension by tying the rope back to itself. The key feature of the rolling hitch is that it locks in place, so it won't slip.

1. Wrap a length of rope around a standing object (a table leg or human leg, for example). Create a half-hitch—a very simple knot—by looping the rope around and tucking it underneath itself (figure 1).

2. Roll the rope around one more time (hence the name) and tuck it under itself a second time (figure 2). Throughout this process, it is important to pinch the half-hitch with one hand while you roll the loose end around the second time so that it doesn't lose its tension. At this point, if you pull it tight, the rolled section gets pinned under itself and won't slip, and you won't need to pinch the knot anymore.

3. Finish the knot with a second half-hitch, but keep it slippery, meaning don't pull it all the way through but leave a loop on one side and the loose

end on the other. This will make it easier to untie later, as all you need to do is pull the loose end (figure 3).

Rolling hitches are often used as part of a system for tightening a rope, where one end, attached with a bowline, is tied to a starting point, say the grommet on a tarp, and then goes around a tree, for example, and back to a bowline on a bight on that same stretch of rope somewhere between the bowline and the tree, where the loose end is then pulled through the loop of the bowline on a bight, which allows you to really crank down on the rope, and is finally tied to itself with a rolling hitch. In a system like this, an important detail is that the rolling hitch itself actually captures the rope on both sides of the tree, pinning them together by wrapping around both at once.

A typical combination of a rolling hitch that uses a bowline on a bight to gain some mechanical advantage and pull down tightly.

If you're confused, picture the bowline tied to the bow of a canoe that you're strapping to the roof of your car. The rope then goes under the front bumper and gets attached somewhere, then comes back up to a bowline on a bight about at windshield wiper level, goes through that loop and is pulled down to tighten, and then is tied off with a rolling hitch. Most people secure things like this with a trucker's hitch, which is a poor substitute for a number of reasons, but you get the idea. This use of a bowline on a bight essentially makes the arrangement a crude pulley system, with tremendous mechanical advantage, and the rolling hitch is crucial to keep the rope from slipping under the force exerted under these circumstances. Untie by pulling on the loose end to undo the half-hitch and then grip the rolled section of the rolling hitch in your fist and shift it a fraction of an inch to loosen.

The Zeppelin Bend

The zeppelin bend is so named because, the story goes, a zeppelin captain once required his aircrew use this knot, and only this knot, when tying two ropes together. This knot looks complicated at first, but it is so symmetrical that it is actually quite easy.

The ability to tie two ropes together so that you can always untie them no matter how tightly they get pulled is an important skill. The most common way people do this is with an overhand knot made in both parts at the same time, which works fine for something flat like a shoelace, but a rope with a round cross section tends to untie itself by rolling right up to the loose ends if it comes under enough strain. If you don't believe me, try it sometime. And while you *could* just attach a bowline to a bowline to tie two ropes together, that method is inferior to the zeppelin bend, the knot I actually use for these circumstances.

1. Form loops with the loose ends of the two ropes you wish to join (for practice, just use two ends of the same piece) so that one forms a lowercase *p* with the loose end on top and the other a lowercase *d* with the loose end on the bottom.
2. Place the *p* on top of the *d* so that the loops in the letters line up (figure 1).
3. Bend each loose end around and through the loop, passing side by side (figure 2).
4. Tighten the knot by holding the loose ends in your thumb and forefingers of each hand, and pulling in opposite directions with the middle, ring, and pinkie fingers. This knot has two backs that can be broken to untie it (see page 230).

The Square Knot

This is a knot people love to hate. Yes, it will eventually loosen, unlike a number of alternatives. Yes, people often tie it incorrectly, making its tendency to loosen even worse. But there are plenty of times when it is exactly what you need, and anything more would be overkill.

A square knot is just two overhand knots in a row, but (and this is *the* key), the overhand knots are tied in opposite directions. So if the first is tied left over right (figure 1), the second is tied right over left. Or vice versa; it doesn't matter so long as you switch whatever you did

for the second one. If you tie two over- hand knots the same way both times, you end up with a granny knot. A granny knot is not a great knot (apologies to everyone who has an awesome granny).

The square knot (figure 2) is not some- thing people tend to tie loose. It is almost always tied under pressure (like how you tie your shoelaces under pressure). The key is to tie the first overhand knot and gain the pressure by pulling it tight (again, think of your shoelaces), pin the first knot with your ring finger and tie the second, reversed overhand knot, then tighten it down and slip your finger out of the way at the last moment. It usually pays to make this second knot slippery, which is a fancy way of saying pull just a loop through from one of the sides, so all you need to do to untie it is pull the loose end. This is the knot to use to tie up bundles of things, or when you forget your belt and are using baling twine instead.

The Slipped Half-Hitch

We already covered the half-hitch as part of the rolling hitch, but it's worth mentioning again because it is so useful in its own right. The slipped half-hitch falls in the same category as the square knot: not bomb-proof, but often it's all that you need.

A slipped half-hitch is ideal when you need to pull something tight and then tie one little knot to hold that tension.

1. First, wrap the rope around a leg or a post (that's the part that holds the tension).
2. Twist it around its doubled-back self, not in a forward spiral, but back toward the object you just wrapped it around so that it tucks under itself and pins itself down (figure 1). That's a half-hitch. Use a finger to pin the rope to the object it wraps around; pinning the rope this way holds the tension while you tie the half-hitch itself. Otherwise as soon as you go to tighten the half-hitch it will loosen up.

3. You could do this as many as four times and it would hold almost any amount of strain. In this case, only do it once, and make it slippery by pulling only one loop of the rope through, not the whole thing (figure 2). Now you only need to pull on the loose end to untie it.

I use the slipped half-hitch as the final knot on a bale of greens. If you want to make it extra secure but still somewhat easy to untie, tie a slipped half-hitch and then an extra half-hitch on top of that with the loop you pulled through. The bulkiness of the loop makes it easier to untie but the extra hitch keeps anything from snagging on the loop and untying the knot. The single slippery half-hitch is not particularly secure but often is all you need. But when tying down a half ton of trees being delivered, for instance, a second half-hitch stacked on top the first is crucial for security.

Half-hitches are a great way to tie off something under tension. Here, I've used them to tie off the end of the rope holding down a load of trees or greens.

While the knot is harder see here, a couple of slipped half-hitches are how every bale of balsam greens gets finished off.

240

Making Your Own Scythe Handle

Because there is more to making these handles than makes sense to dive into given the scope of this book, I will restrict myself to some of the most helpful tips. Think of making your own scythe handle as a journey; following these tips will get you a great deal of the way there, but not necessarily to the finish line. Your second handle may (or may not) be better than the first, and the third hopefully better still. Working with wild materials like raw, unfinished saplings requires the ability to see their important underlying geometry and ignore all the wiggles in between. The process of making a handle is imprecise for a tool that ultimately requires precision. The only way to achieve this is through experience. However, keeping these general guidelines in mind will help.

Use a bendy sapling. While you could make a handle from a straight piece, one with bends will often work better. Once you understand which bends and dimensions are important, you can easily scrutinize a sapling and decide if it's worth harvesting, if it needs another year or two of growth, or if it's too big.

Measure the diameter. It will save you a lot of time if the section of sapling you intend to take (generally a 5- or 6-foot length/1.5 to 1.8 m) is 1½ inches (3.8 cm) in diameter at one end and 2–3 inches

(5–7.5 cm) at the other. The best place to find saplings like this is along the edge of the road, where saplings tend to bend outward to reach the light and then back upright when they find it, often achieving the correct sort of curves in the process.

Get permission. Roads also make it convenient to scout a large area, but get permission from the landowner before cutting on their land. In very rural areas, and throughout New England where trees are plentiful, keep in mind that the town technically owns 5 feet (1.5 m) from the road edge, so on small dirt roads this leaves a lot of saplings within the town's management. Even so, if pursuing this form of public harvesting, find an area where it's not going to be missed. Around me, road edges where I can't see houses, and where the town is using a flail mower to keep the saplings down, seem like fair game. I also cultivate saplings specifically to use as scythe handles in the balsam grove.

Choose your wood. Any species of wood will work for the handle, although it's worth shying away from the heavier woods such as hickory or oak. The first choice is probably ash, followed by maple and birch. Even light, relatively weak woods such as poplar and willow should work fine as long as you use a thicker piece.

Get a grip. The scythe handle's grips are made from tree crotches, and finding the right angle is crucial. Ideally the angle between the two arms should be somewhere between 45 and 90 degrees. The best place to find such a fork is high up in the branches, so finding a tree that has been taken down is ideal; any treetop should offer several good pieces. The best species for finding the correct angles and proportions for grips tend to be birch, beech, and cherry. Don't use poplar, even though its angles are often ideal, because the joints are weak and will rip apart under the stresses of use.

Keep your options open. It's a good idea to keep a bunch of options on hand for handles and grips, since you never know what combination will work best. Also, make the grips several months before harvesting the handle, so they have a chance to dry and shrink before you install them. If you insert green, fresh grips into a seasoned handle they will shrink and loosen over time.

Having an assortment of potential saplings can make choosing a handle easier. *Photo by Joshua Klein.*

Grips are the hardest part to find, particularly the right-hand ones. I harvest them whenever I can. *Photo by Lauren Bruns.*

Carve scythe grips down with an axe and knife to be the shape of a stick of butter squeezed in your hand. *Photo by Joshua Klein.*

Build the handle. Build the handle by putting the blade on the ground and holding the sapling as if it was already the handle (figure 1). Make sure the sapling's thin end is positioned at the blade so that the thick end up top can balance out the weight of the blade. Turn the sapling this way and that to figure out the right orientation, which will determine where to cut the tang face and the tooth hole for the blade to clamp onto, and where to place the grips. Cut a flat plane to support the tang (figure 2), and then cut a mortise for the tang tooth by drilling a hole and carefully trimming it square with the knife tip (figure 3). Finally, shave down the end of the handle until the ring clamp is able to slip down past the tang face (figure 4), then attach the blade to the handle with the ring clamp before determining which grips to use (figure 5). Grips should be fitted where your hands naturally fall when dangling at your sides with your back straight and your knees slightly bent. Figure out the precise placement of where the grips should go by holding the sapling in different places and paying attention to which positions feel the best and require the least effort to swing the blade.

Secure the grips. Drill holes with a ½- or ¾-inch spade bit (depending on the size of the grip stem), and remember the hole needs to match the direction of the grip stem, which might not be straight! When you're not using

244

machined parts, you need to be attentive to the actual angle of the wood, and adjust what you are doing to that. I drill using a spade bit, and stop when the pilot tooth just starts to poke through the back side of the handle. I then trim the length of the grip stem to place the grip where I want it in relation to my hands hanging at my sides, taking into account the depth of this mortise hole. Shave the stem down to the right size to fit snugly in the hole, checking often so that you don't overcut it. Use wood glue or five-minute epoxy to secure the grips. I also use a short nail hammered in from the side of the handle (I drill a pilot hole first) to further secure the grips in the handle.

Oil it. No finish is necessary, but if you want, it's fine once the handle has dried to brush the whole thing down with linseed oil cut with turpentine, or with tung oil. It usually takes several months for the handle to dry if the bark is still on. I used to remove the bark, but then it seemed like an unnecessary step so I stopped.

This sequence of photos by Joshua Klein.

With longer blades, it can be helpful to choose a sapling that bends such that a line drawn from one grip to the next and continued on to the blade would hit the thickest part of the blade, which has the effect of balancing the blade's weight from side to side. This helps keep the blade from feeling tip-heavy, reducing fatigue and minimizing the chance that the tip will catch on the ground.

RESOURCES

Farming and Gardening

Four Season Harvest: Organic Vegetables from Your Home Garden All Year Long, by Eliot Coleman (Chelsea Green Publishing, 1999). The classic book that lays the groundwork for all winter vegetable production in hoophouses.

The Intelligent Gardener: Growing Nutrient Dense Food, by Steve Solomon and Erica Reinheimer (New Society Publishers, 2012). A game-changing book for me, forever altering how I understand soil and plant and human health. A must read.

The Lean Farm: How to Minimize Waste, Increase Efficiency, and Maximize Value and Profits with Less Work, by Ben Hartman (Chelsea Green Publishing, 2015). The best articulation I've come across about how to analyze the systems in a farm or business and determine how to improve.

The Winter Harvest Handbook: Year-Round Vegetable Production Using Deep-Organic Techniques and Unheated Greenhouses, by Eliot Coleman (Chelsea Green Publishing, 2009). This one takes Eliot's experiments to the next level of refinement, locking in systems that work for growing food through the winter using hoophouses.

Scythe Connection (www.scytheconnection.com). Peter Vido's website on scything is *the* source for European-style scythes. Videos, technical writing, it's got it all.

The Bionutrient Food Association (www.bionutrient.org). Started by Dan Kittredge, this is the most prominent organization I know of promoting the use of remineralization to improve soil, and plant and human health.

Structures and Landscape

A Pattern Language: Towns, Buildings, and Construction, by Christopher Alexander, Sara Ishikawa, and Murray Silverstein

(Oxford University Press, 1977). Like the Bible, you can pick this up and always find something new. This one is constantly in play in my home.

The Arts of the Sailor: Knotting, Splicing and Ropework, by Hervey Garrett Smith (Dover Maritime, 2012). A classic book that covers all the knots as well as canvas and ropework of all kinds. A must if you want to learn to use rope and canvas tarps.

The Small House Book, by Jay Shafer (The Tumbleweed Tiny House Company, 2010). There are a ton of tiny house books out now. This is the only one I have experience with, and I would recommend it to anyone planning to build a structure on wheels simply for the detailed information about attaching a structure to a trailer frame.

Tiny Tiny Houses: Or How to Get Away from It All, by Lester Walker (Overlook Books, 1987). This is the book for the dreaming phase of any tiny build.

Spoon Carving

Spoon: A Guide to Spoon Carving and the New Wood Culture, by Barn the Spoon (Scribner, 2017). Barn's book is one of the most evocative articulations of why spoons and spoon carving matter.

Spoon Carving, by EJ Osborne (Quadrille Publishing, 2017). EJ's ability to embrace the wonkiness inherent within spoon carving and frame the craft within a narrative of self-expression makes this book the perfect place to start.

The Art of Whittling: A Woodcarver's Guide to Making Things by Hand, by Niklas Karlsson (Carlton Books, 2018). This might be my favorite book about spoon carving. Niklas has an uncanny ability to couch the process and technical details within his Sami heritage, which makes the whole thing more approachable.

Robin Wood (www.robin-wood.co.uk/wood-craft-blog). Robin's blog is a treasure trove and, while no longer representative of the current state of the field, was where I gained much of my initial information.

Spoonesaurus (www.spoonesaurus.com). The Spoonesaurus Instagram account, which can be linked to from the main website, is a solid compilation of nuanced one-minute videos about spoon carving. A great place to go once you've dipped your toes in.

Business

Perennial Seller: The Art of Making and Marketing Work That Lasts, by Ryan Holiday (Portfolio, 2017). This book has proven invaluable for my education on how to think about building a community that supports my work. Even if you are not a writer it is worth reading, preferably several times over.

The Thank You Economy, by Gary Vaynerchuk (HarperBusiness, 2011). Also see his other books, podcast, social media, and You-Tube channel. If you are only going to follow up on one resource, start with this guy. Go to YouTube and find his video titled *How to Start*. Gary's is the voice I hear in my head all day long, both because I listen to his podcast but also because his articulation of empathy, kindness, hard work, hustle, and strategic business thinking has become my inner voice. I wouldn't be where I am without him.

Seth Godin (https://seths.blog). Seth's daily blog is a potent reminder of the things that matter in business and in life. I am so grateful it has come into my life. He has also written many books, and has an excellent podcast called *Akimbo*.

INDEX

Note: Page numbers in *italics* refer to figures and photographs.

ABOUT THE AUTHOR

Emmet Van Driesche, along with his wife, Cecilia, operates the Pieropan Christmas Tree Farm in western Massachusetts. When he's not working on the tree farm or editing scientific manuscripts, he spends his time carving wooden spoons and teaching others to do the same. You can find out more about him at www.emmetvandriesche.com.

ABOUT THE AUTHOR

Emmet Van Driesche, along with his wife, Cecilia, operates the Pieropan Christmas Tree Farm in western Massachusetts. When he's not working on the tree farm or editing scientific manuscripts, he spends his time carving wooden spoons and teaching others to do the same. You can find out more about him at www.emmetvandriesche.com.

the politics and practice of sustainable living

CHELSEA GREEN PUBLISHING

Chelsea Green Publishing sees books as tools for effecting cultural change and seeks to empower citizens to participate in reclaiming our global commons and become its impassioned stewards. If you enjoyed reading *Carving Out a Living on the Land*, please consider these other great books related to agriculture and business.

UNCULTIVATED
Wild Apples, Real Cider, and the Complicated Art of Making a Living
ANDY BRENNAN
9781603588447
Hardcover • $24.95

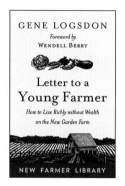

LETTER TO A YOUNG FARMER
How to Live Richly without Wealth on the New Garden Farm
GENE LOGSDON
9781603588065
Paperback • $18.00

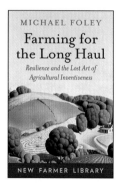

FARMING FOR THE LONG HAUL
Resilience and the Lost Art of Agricultural Inventiveness
MICHAEL FOLEY
9781603588003
Paperback • $20.00

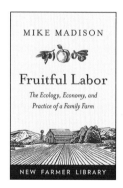

FRUITFUL LABOR
The Ecology, Economy, and Practice of a Family Farm
MIKE MADISON
9781603587945
Paperback • $18.00

For more information or to request a catalog,
visit **www.chelseagreen.com** or
call toll-free **(800) 639-4099**.

the politics and practice of sustainable living

CHELSEA GREEN PUBLISHING

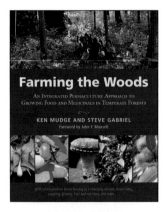

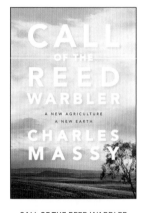

FARMING THE WOODS
*An Integrated Permaculture Approach to
Growing Food and Medicinals in Temperate Forests*
KEN MUDGE and STEVE GABRIEL
9781603585071
Paperback • $39.95

CALL OF THE REED WARBLER
A New Agriculture, A New Earth
CHARLES MASSY
9781603588133
Paperback • $24.95

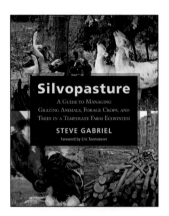

SILVOPASTURE
*A Guide to Managing Grazing Animals, Forage Crops,
and Trees in a Temperate Farm Ecosystem*
STEVE GABRIEL
9781603587310
Paperback • $39.95

TREES OF POWER
Ten Essential Arboreal Allies
AKIVA SILVER
9781603588416
Paperback • $24.95

CHELSEA
GREEN
PUBLISHING

the politics and practice of sustainable living

For more information or to request a catalog,
visit **www.chelseagreen.com** or
call toll-free **(800) 639-4099**.